PRACTICAL OSCILLOSCOPE HANDBOOK

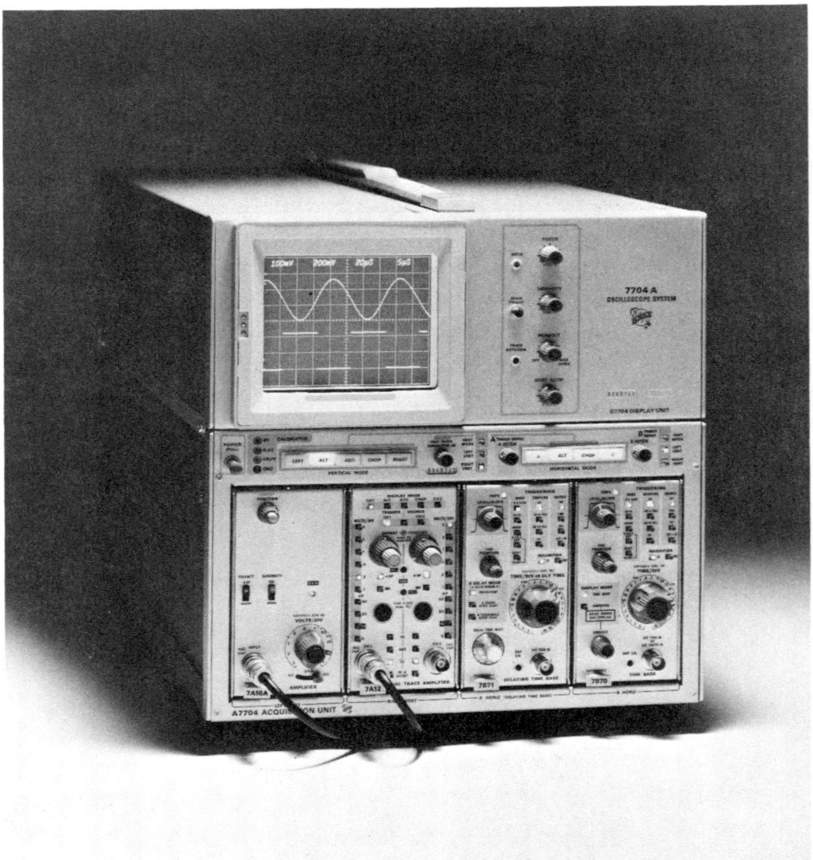

More than just an oscilloscope: a modern laboratory instrument with four plug-ins.

PRACTICAL OSCILLOSCOPE HANDBOOK

John Douglas-Young

PARKER PUBLISHING COMPANY, INC.
WEST NYACK, NEW YORK

Parker Publishing Company, Inc.
West Nyack, New York

Library of Congress Cataloging in Publication Data

Douglas-Young, John.
Practical oscilloscope handbook.

Includes index.
1. Cathode ray oscilloscope--Handbooks, manuals, etc.
I. Title.
TK7878.7.D68 621.3815'48 78-21522
ISBN 0-13-693549-4

Printed in the United States of America

How This Book Will Help You

Oscilloscopes come in all prices, from under $100 to over $8000, and in all capabilities, from relatively simple to incredibly sophisticated. This book will help to greatly increase this instrument's value to you, for in skilled hands the oscilloscope is the most powerful of all troubleshooting tools, and truly indispensable for professional work in today's electronics.

Every chapter in this handbook gives you practical, essential guidance in the key aspects of selection, operation and maintenance of your scope; information that you previously would have had to search for through several volumes. It begins by explaining how your instrument's performance and all-round usefulness depend upon five major features: frequency response, input impedance, sensitivity, sweep type and method of synchronization. Knowing these, you will be able to tell whether the scope you have, or are considering acquiring, is suitable for your purpose.

Some people have only hazy ideas about the functions of many of the controls on the scope's front panel. They learn by rote how to set those necessary to obtain a routine display and leave the others alone. A considerable portion of your scope's efficiency is of no benefit if it is not used. This book describes the uses of *all* controls, and points out the proper methods of making connections for best results. By following these directions, your scope will rapidly become as easy for you to handle as your automobile, no matter how complex it may appear at first.

Understanding the front panel is greatly facilitated by understanding what goes on behind it. The amazing circuitry of an oscilloscope is a marvel of ingenuity. It is capable of measuring a very wide variety of rapidly changing electrical phenomena, even those that happen only once and last for only a fraction of a microsecond. This book covers it all, from the probe to the cathode-ray tube, in a clear, non-technical, abundantly illustrated format.

This alone would be enough for a complete book by itself, but there is much more. Whole chapters on *troubleshooting* with your scope follow. You are shown how to use it skillfully as a signal tracer to trace the signal through the TV receiver. By using this technique you can quickly locate a dead stage that the signal will not pass through at all, or a defective stage in which the signal is being distorted. (This same procedure works for all types of electronic equipment, of course.)

You are also shown how to check the response of the video i.f. amplifier using sweep and marker generators, and how to make adjustments, if necessary. Other types of amplifiers may be tested similarly, using a square-wave generator to inject a signal for display on your scope screen so that you can see if the amplifier is reproducing it faithfully or is distorting it.

A knowledge of waveform analysis is a great help in troubleshooting, and a complete chapter is devoted to this. All the principal waveshapes are dissected and discussed. You are shown what deviations from the ideal shape mean, and how to measure the frequency and amplitude of the fundamental and harmonic components.

This book will reveal your scope to you as the most versatile, informative and interesting instrument in your shop, and you will enjoy a satisfying sense of accomplishment in having mastered this extraordinarily helpful instrument. By applying the procedures in this book you will also save hours of troubleshooting time, which means more profit, prestige and just plain pleasure in being able to use with skill and confidence one of the most powerful tools on the electronic technician's workbench.

John Douglas-Young

CONTENTS

LIST OF ILLUSTRATIONS

Other Books by John Douglas-Young

Complete Guide to Reading Schematic Diagrams

Complete Guide to Electronic Test Equipment and Troubleshooting Techniques

Technician's Guide to Microelectronics

1

Understanding Oscilloscopes

An oscilloscope is for looking at waves . . . that are actually "invisible" surges of electrons. Since all electronic equipment functions by the action of electrons, the behavior of electrons in electronic circuits reveals how they are working. The oscilloscope makes the invisible waves visible, so we can actually "see" the inside operation of electronic equipment.

By using transducers (microphones, light-sensitive cells, and so on), we can also produce electrical waves that correspond in form to physical phenomena. In this way we can use the oscilloscope to study non-electrical functions, such as sound waves, heart beats, engine performance.

When we connect a test lead between the input terminal of our scope and the point in the circuit we wish to examine, some of the electrons surging back and forth in the circuit are diverted along the lead to the scope, where they detour through a one-megohm resistor and come back to the circuit via the ground-return path. As a surge increases, a potential builds up across this resistor; as a surge decreases, the potential diminishes; as it reverses direction, the potential changes from positive to negative, or negative to positive, as the case may be.

After amplification in the scope circuits, this varying potential traces a line on the scope screen that moves upward or downward in accordance with its value from instant to instant. In a given period it may do this several times.

To show these excursions in proper sequence the trace is also made to travel sideways at a steady rate corresponding to the time period, so that the up-and-down excursions are drawn out into an undulating line across the screen, as in Figure 1.1.

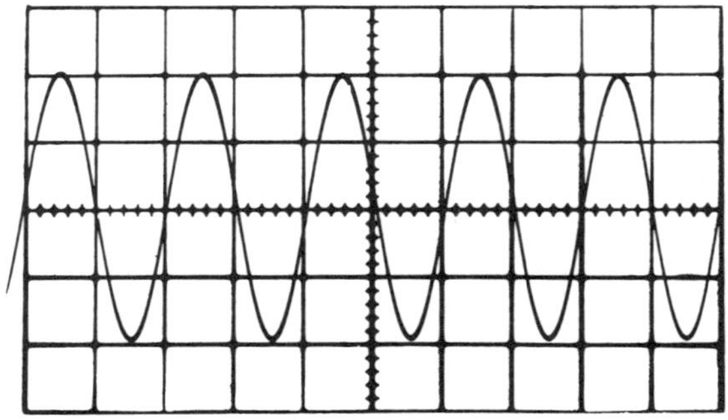

Figure 1.1 A waveform is displayed on your scope screen like a graph.

The shape of this undulating line is called a *waveform*. It consists of a very large number of instantaneous values of the varying potential plotted against time, in rectangular coordinates as in a graph. In electronics, currents that vary periodically with time are commonplace, and appear on the scope screen as a series of evenly spaced waves. From this display you can evaluate as many as twenty different electrical characteristics of the waveform.

Many other instruments exist that are used to measure electrical characteristics, but each is designed to measure one characteristic only. A current meter, for instance, measures current, but unless the current varies at a very slow rate, which is seldom the case, it can only show the average value, not individual excursions. The same applies to a voltmeter. Similarly, a counter can only count peaks, positive or negative, and a distortion meter can only measure harmonic distortion.

The difference between these instruments and an oscilloscope is like the story of three blind men who encountered an elephant. One of them bumped into a leg, which to his sense of touch felt like a tree trunk. The second brushed against the pachyderm's tail, which he deduced was a dangling piece of rope. The third collided with Jumbo's proboscis, which the friendly beast twined around his waist, almost scaring him to death. After discussing the matter among themselves, they concluded the object was a tree, from which hung a rope and a boa constrictor. Of course, if they had been able to see the whole animal instead of having only their sense of touch to go by, they would have recognized it for what it was. The oscilloscope allows us to see "the whole animal" as no other instrument can.

There are a great many different kinds of oscilloscope, from the simplest service scope to the advanced modular laboratory instrument, but they all have one thing in common: the cathode-ray tube. The first CRT was invented in 1879, eighteen years before the scope itself, by Sir William Crookes. He used it to study "cathode rays." The Crookes tube had two internal metal electrodes, and contained gas at very low pressure. When a high voltage was applied between the electrodes, a mysterious beam appeared that glowed in the dark. It seemed to emanate from the electrode with the negative charge (the cathode) and stream toward the electrode with the positive charge (the anode). The color of the glow depended upon the gas used.

What made cathode rays mysterious to the scientists in 1879 was that they did not know of the existence of electrons. It was the ionization of the gas by a beam of electrons that caused the glow, but it was not until 1897 that Sir Joseph Thomson discovered that a subatomic particle was responsible. He demonstrated its existence by introducing a fluorescent screen and deflection plates into the tube. By applying various potentials to the deflection plates he not only deflected the beam of "corpuscles," as he called them, but was able to show that an electron was 1860 times lighter than a hydrogen atom. In that same year Karl Friedrich Braun constructed the first oscilloscope, using a similar tube, to investigate the properties of high-frequency alternating currents.

None of these tubes had a heated cathode, so potentials as high as 50 kilovolts were required to produce an electron beam. Such a beam is very "stiff," and takes very high deflection voltages to bend it. In 1921, J.B. Johnson designed a new tube with a heated cathode that needed only 300 volts to produce an electron beam. This tube was essentially the same as the modern CRT described in Chapter 2.

Another "descendant" of the original Crookes tube is the television picture tube. Because both the CRT and the picture tube are cousins, as it were, the general organization of the scope and the TV receiver shows a definite family resemblance. In Figure 1.2 you can see that both have a main signal path from the antenna or probe to the picture tube or CRT as the case may be. Each also has deflection circuits. The TV receiver has audio and color sections that are absent in the scope, while the scope has a calibrator not found in the

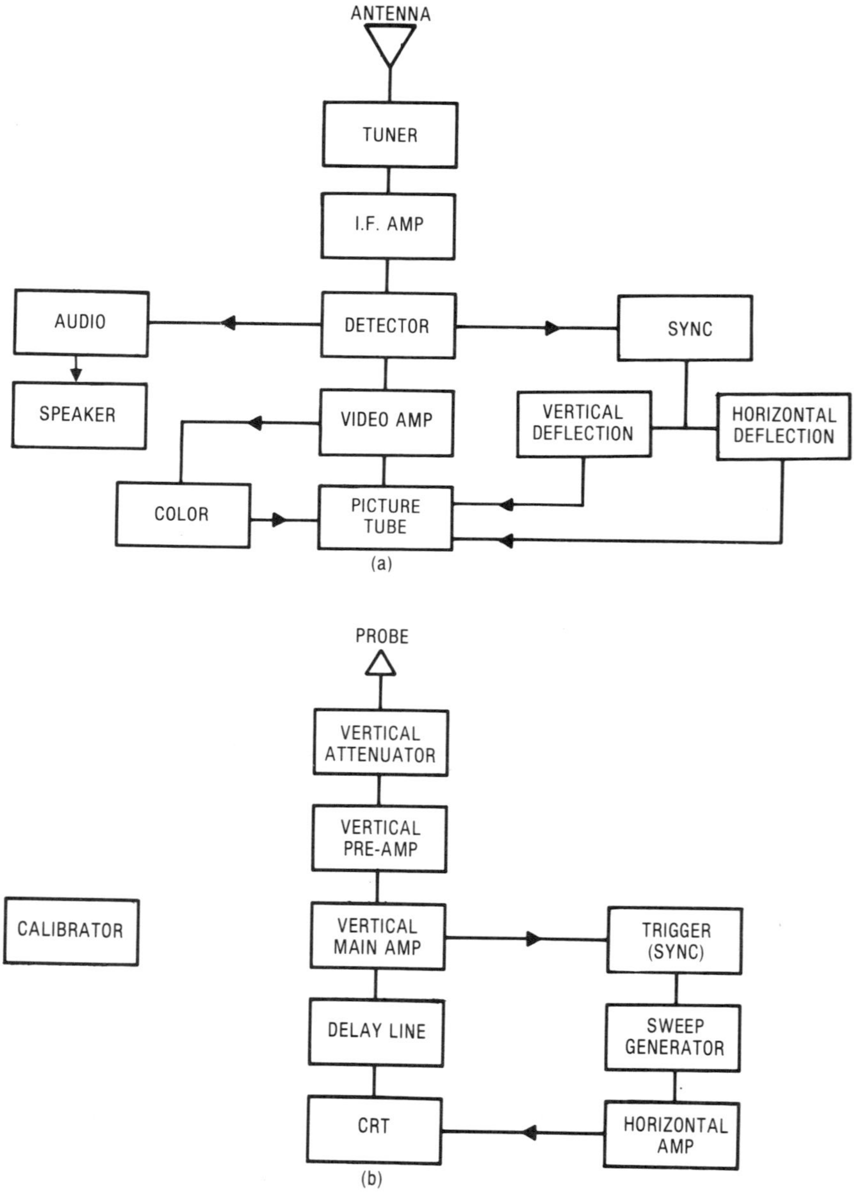

Figure 1.2 (a) TV Receiver (12 sections). (b) Oscilloscope (10 sections) (Power Supplies omitted—they are similar)

TV receiver. Not shown are the low-voltage and high-voltage power supplies, which are very similar in both cases.

Ignoring the audio and color sections, which have no counterparts in a scope, the main difference between them is that the electron beam in the picture tube is deflected vertically as well as horizontally. After each horizontal sweep, the following sweep traces the next line slightly lower, until the whole screen has been covered. This process is repeated 30 times a second, so all the 525 horizontal lines appear stationary on the screen, if you are close enough to see them. The picture information is applied between the picture-tube grid and cathode so that it varies the electron flow, and therefore the intensity of the beam, to produce different degrees of brightness, and so form the light and dark areas of the picture. (This becomes a little more complex in a color set, but the principle is the same.)

But in a scope there is a horizontal sweep only, so there can be just one horizontal line. And instead of varying the brightness of the trace according to its amplitude the input signal moves it up and down.

The picture tube has another important difference. Since the TV horizontal and vertical deflection circuits run at constant rates, the electron beam is deflected by the magnetic fields induced by currents in external coils. These have the advantage of bending the beam through a greater angle, so the ratio between screen width and tube length is less than in a CRT. However, the electrostatic deflection produced by the potentials on the internal deflection plates in the CRT are equally linear regardless of sweep rate, signal frequency and so on, and are therefore the only form of deflection that will work equally well over the entire bandwidth of the scope.

Figure 1.3 shows the front panel of a typical medium-priced scope with a bandwidth from d.c. to 15 megahertz. This instrument has characteristics that make it suitable for the service shop, the production line, the classroom, even the laboratory, though it lacks many of the refinements of the more advanced modular models. It does have two separate input channels so you can display two signals at once, as you can see on the screen in the figure.

As you examine the controls of this instrument you will realize at once that very few of them duplicate those of a TV receiver. There are, of course, no controls for sound or color, and only the power and intensity (brightness) perform the same function in a

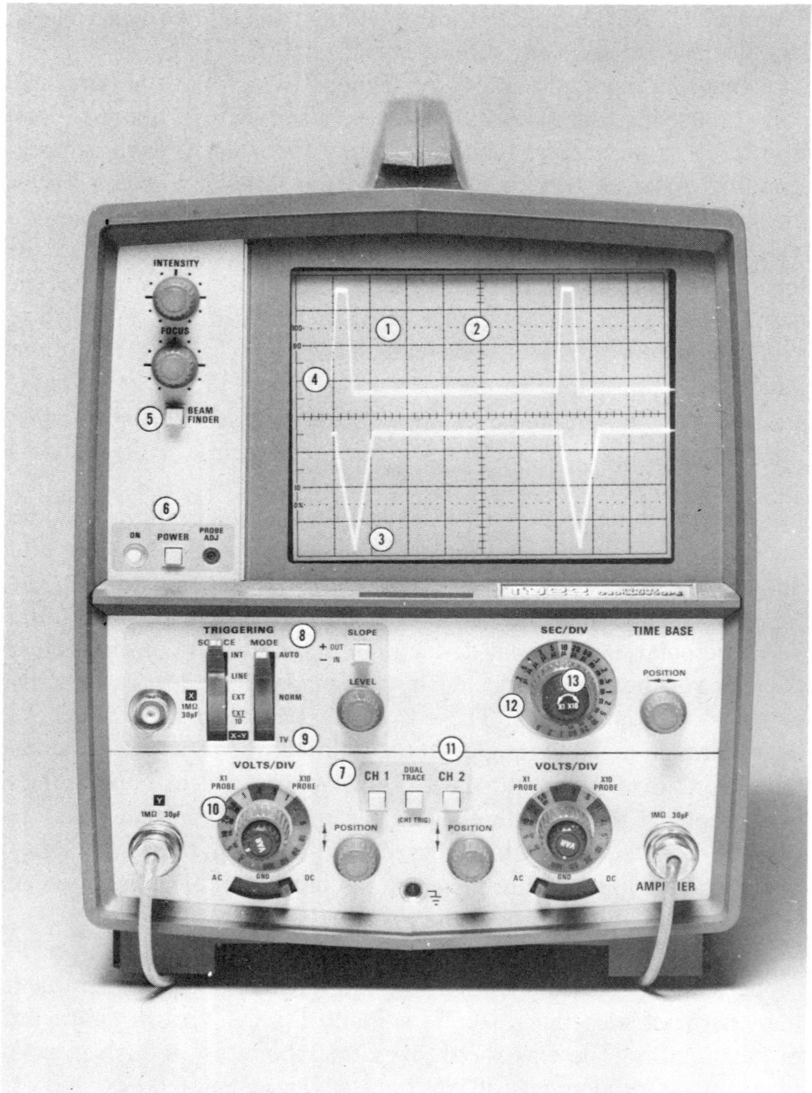

Figure 1.3 Front panel of modern general-purpose scope showing typical controls. (*Reproduced by permission of Tektronix, Inc. All rights reserved.*)

---

1. **12kV, post-deflection-accelerator crt.** Provides bright 8 x 10 cm displays of low-rep-rate signals, even under adverse lighting conditions.

2. **Internal graticule.** Helps eliminate parallax errors in critical measurements.

3. **Small, clearly-defined spot.** Ensures sharp traces and high resolution.

4. Built-in delay-line. Let's you see the leading edge of triggering signal.
5. Beam-finder. Quickly reveals presence or absence of input signal and locates the trace position.
6. Power ON light. Flashes if line voltage exceeds power supply regulation limits. (Regulated power supplies help insure 3% accuracy in amplitude and timing measurements.)
7. Color-coded controls. Shorten familiarization time and aid in easy operation. All functionally related controls are colored the same, and grouped for greatest convenience. Green is reserved for triggering controls, blue for mode selectors and red for variable (uncalibrated) controls.
8. Automatic trigger mode. Provides free-running internal trigger in absence of input signal for quick location of sweep trace.
9. TV trigger mode. Automatically provides line or frame-triggered displays for best viewing of composite tv signals.
10. Deflection factor switch. Wide range (2mV to 10V/cm) accommodates all commonly encountered signal amplitudes. Clear windows on switch skirt yield two readings—one for X1 probe, the other for X10 probe. No distracting mental arithmetic to perform. Compensated step attenuator offers constant input impedance across entire bandwidth.
11. Display mode switching. Dual-trace mode provides automatic selection of alternate or chopped-sweep displays for best viewing, depending on selected sweep rate.
12. Sweep rate selector. Provides sweep rates from 200 ns to 0.5 s/div in 18 steps. Clear window on switch skirt clearly identifies sweep time/cm.
13. Sweep expander control. Variable from X1 to X10. Calibrated in X10 detent position for 5% accuracy. Expands center one centimeter of display by factor of ten (max), providing maximum equivalent sweep rate of 20 ns/div. For detailed displays of selected waveforms, just position signal to center of screen and turn sweep expander to desired magnification level.

---

scope as in the TV set. The panel is divided into three sections. The lowest is the *amplifier* panel with two inputs, two attenuator switches (designated VOLTS/DIV), two position controls that move the traces up and down independently, two input-mode switches and three push buttons with which you select either channel 1, channel 2, or both, as shown.

Above the amplifier panel is the *time-base* panel with the time-base or sweep-rate selector control (designated SEC/DIV). There is also a position control for adjusting the horizontal position of the trace. Then there is a group of controls in a colored area marked TRIGGERING. These controls are all concerned with synchronizing the sweep with the input signal to obtain a stable display.

The third section of the panel, to the left of the screen, has three controls associated with the CRT, the power on-off switch and an output signal for adjusting the compensated probes supplied with the scope.

The purpose and operation of these controls will be described fully in the next four chapters as we describe the circuits they operate in.

Main Types of Scope

Scopes can be divided into two main classes: *monolithic* and *modular*. The monolithic scope is a one-piece instrument designed for a single range of functions. This makes it less versatile, but it is generally smaller, lighter and less expensive. Service scopes, portable scopes, patient monitors, TV vectorscopes and spectrum analyzers are examples of monolithic scopes. Because they are restricted to specialized tasks they are also called "dedicated."

Modular instruments combine a *mainframe* and one or more interchangeable *plug-in* subassemblies. They are often called "plug-in" or "laboratory" scopes. By using various plug-ins many more functions can be made available than could be practically combined in a monolithic instrument, not to mention the cost and bulk of such a scope if you could get one. A modular scope can also be kept "modern" with new plug-ins as they come out, whereas a monolithic instrument can only be modernized by replacing the whole thing. In the most advanced modular scopes the mainframe contains the CRT, power supplies and switching circuits only. The vertical amplifiers, time bases and so on, are all plug-in assemblies.

Whatever their class, the majority of scopes are designed to display the waveform of a signal as a graph of the signal's amplitude against time. In these instruments the vertical amplifier, or amplifiers, is the primary section, with every possible refinement included to ensure the best fidelity over the entire bandwith. The time-base section and triggering also are superior to enable full advantage to be taken of the vertical amplifier's capabilities. We shall discuss these circuits in detail in the following chapters.

There are three types of waveform-display scopes: the type with a conventional CRT, the type with a storage CRT, and the sampling oscilloscope. Storage CRT's have the ability to retain and display the image of an electrical waveform on their screen after the waveform ceases to exist. The image may be retained for a few seconds to several hours, and may be erased when no longer required.

Even though signals in the gigahertz range are well beyond the bandwidth limits of conventional scopes, you can still display them by a technique called *sampling*. The manufacturers of modular scopes make sampling plug-ins that convert these instruments to sampling scopes.

Sampling is a technique that displays the amplitude *at one instant only* of each cycle of a recurrent waveform. Each point displayed is slightly later in time than the one next before it, so eventually "samples" will have been taken of every point on the waveform, as shown in Figure 1.4(a). Each sample appears as a dot

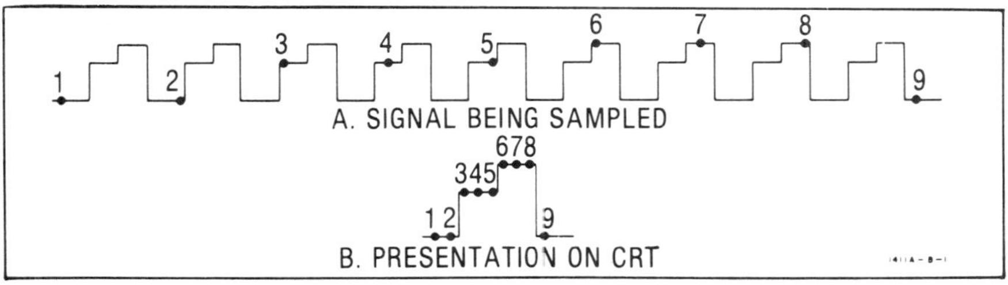

Figure 1.4 (A) Samples are taken from successive cycles, each sample coming from a point slightly further along on the cycle than its predecessor. (B) When all the samples are displayed close together, they show one complete cycle. Up to 1000 samples are taken.

on the CRT screen, but as the sweep rate is very much slower than the signal frequency the dots are close together, and have the appearance of a continuous trace. In Figure 1.4(b) there are only a few dots, so you can see how this works, but in a real display there will be as many as a thousand.

Nearly all scopes provide a "horizontal" or "X" input. In Figure 1.3 you can see this at the left-hand side of the TIME BASE section of the control panel. The SOURCE switch next to it has a position designated "X-Y". When the switch is placed in this position the sweep generator is disabled and the X input is connected to the horizontal amplifier. As a result the electron beam will now be deflected laterally by any signal applied to the X input, in the same way it is deflected vertically by a signal applied to the Y (or vertical) input.

As we shall see later (in Chapter 5), this mode of operation is used for displaying the phase relationship existing between two signals of the same frequency, or for determining the frequency ratio between two signals, where the ratio is a simple one such as 2:1, 3:1, and so on. A typical use would be to check the phase shift occurring between the input and output of an amplifier. If the input signal is connected to the Y input, and the output signal to the X input, an elliptical figure is displayed, the degree of ellipticity corresponding to the phase angle. Such a display is called a Lissajous figure.

In conventional scopes the Y amplifier is more complex than the X amplifier. It usually includes a delay line as well. This causes a small internal phase shift of up to 5 degrees within the scope itself. For many operations this can be ignored or allowed for, but for really accurate work you need a scope with identical vertical and horizontal amplifiers and no delay line. Such an instrument is called an *X - Y oscilloscope*.

The difference between a *spectrum analyzer* and a conventional oscilloscope or plug-in is that it displays a graph of amplitude against frequency instead of amplitude against time. Basically, it is a tuned receiver, with selectable frequency ranges and spans, coupled to a CRT. With this capability it can display the center frequency and sidebands of a modulated signal. You can also use a spectrum analyzer to measure distortion, signal-to-noise ratio, color-TV signal components, and so on. In the past, spectrum analyzers were very expensive, monolithic instruments. In order to avoid the costly investment required, many specialized tools were developed. These did one or more of the things a spectrum analyzer does, but the latter is many times faster and more efficient. However, today spectrum-analyzer plug-ins are generally used. These are made for operation in modular scopes that perform other functions as well, so are not so costly. As a result, the spectrum analyzer is becoming more popular as engineers and technicians rediscover its power, especially in servicing communications and audio equipment.

Scopes of recent manufacture are, without exception, solid state, with discrete components mounted on printed-circuit boards. Some advanced models also have integrated circuits to provide the logic for alphanumeric data presented on the CRT along with the normal display.* However, many excellent instruments with vacuum tubes are still in daily use, scope manufacturers apparently never having heard of built-in obsolescence. Not very long ago some of these vacuum-tube scopes themselves were leaders in the state of the art, so if you should come across one in good working order offered for sale at a reasonable price, you should give it serious consideration. It may be as good or better than some newer models with much higher price tags.

All the basic circuits in today's scopes were evolved in the vacuum-tube era. Their solid-state versions are almost the same, except that transistors have been substituted for tubes. Some concessions have had to be made, of course, to compensate for the different characteristics of transistors, but the operating principles are identical. Since both versions are in use at present, we shall cover vacuum-tube as well as transistor circuits in the following chapters.

---

*It is, of course, likely that IC's will soon be used in other circuits.

2

Focusing on the CRT

We're starting off with the CRT (cathode-ray tube) because it's the central part of every scope. Like the picture tube in our television set, it's the thing we watch. The rest of the circuits exist only to put the picture on the screen.

The two tubes have much in common. Both are funnel-shaped vacuum tubes with indirectly heated cathodes that emit electrons. Both have anodes to which the electrons are attracted, and grids that control their flow. Both have phosphor screens that glow where the electrons strike. And both use high voltages that can give you a nasty shock!

In fact, they're both CRT's, but for the sake of clarity we'll call the TV tube the picture tube and the scope tube the CRT.

In a black-and-white picture tube a set of deflection coils around the neck keeps the electron beam sweeping over the screen at a fixed rate. The beam covers the screen in successive parallel lines so that the whole area is scanned thirty times a second. Picture details are produced by varying the intensity of the beam (and therefore the brightness at each point) in accordance with the picture information in the TV signal. (Most color tubes have three beams, but the principle is the same.)

In the CRT, on the other hand, a pair of deflection plates inside its envelope sweeps the beam horizontally at whatever rate you select. The input signal is applied to a second pair of plates that deflects the beam vertically in proportion to the amplitude of the signal during the time the beam is crossing the screen. If this fluctuates, as in the case of an a.c. signal, the beam will trace its *waveform*.

In short, the picture tube uses the input signal to *modulate* the electron beam, but the CRT uses it to *deflect* the beam up and down.

There are two main types of CRT: the *conventional* and the *storage*. This chapter covers the following subjects:

- Conventional CRT
  - Screen
  - Electron gun
  - Deflection
- Storage CRT
  - Screen
  - Flood guns
  - Storage modes
- Troubleshooting and replacing the CRT
- Power supplies

Conventional CRT

The conventional CRT is the non-storage tube provided in most scopes. As shown in Figure 2.1, it consists of a narrow neck containing the electron gun and a flared portion with a flat face. (On some older or less-expensive CRT's this face may bulge a little.) The flat face is the screen.

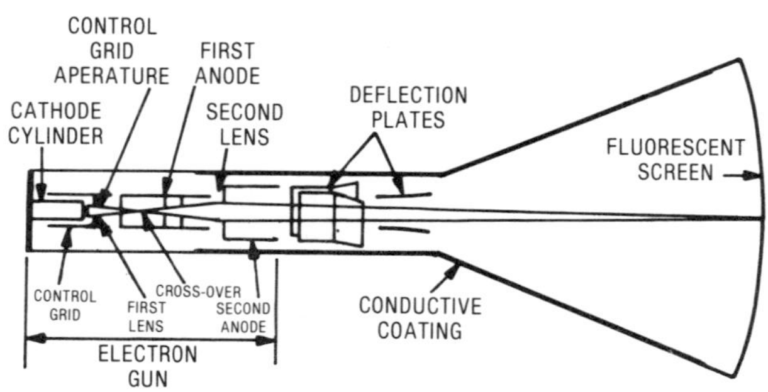

Figure 2.1 Conventional CRT

The screen may be circular or rectangular. At first you could get only circular ones, but eventually the scope manufacturers found out how to make corners on their tubes. As you can see in Figure 2.2, the rectangular tube doesn't waste screen space in the way the circular one does.

The screen is formed by coating the inside of the glass with a *phosphor*. This is a substance that can convert electron energy to light. Electrons are almost weightless (each weighs about 9.1 x 10^{-28} gram), so they can easily be accelerated to very high speeds in the CRT. When they strike the phosphor they are traveling thousands of times "faster than a speeding bullet."

The phosphor is a crystalline substance to which a small amount of an activator has been added. For instance, the P1 phos-

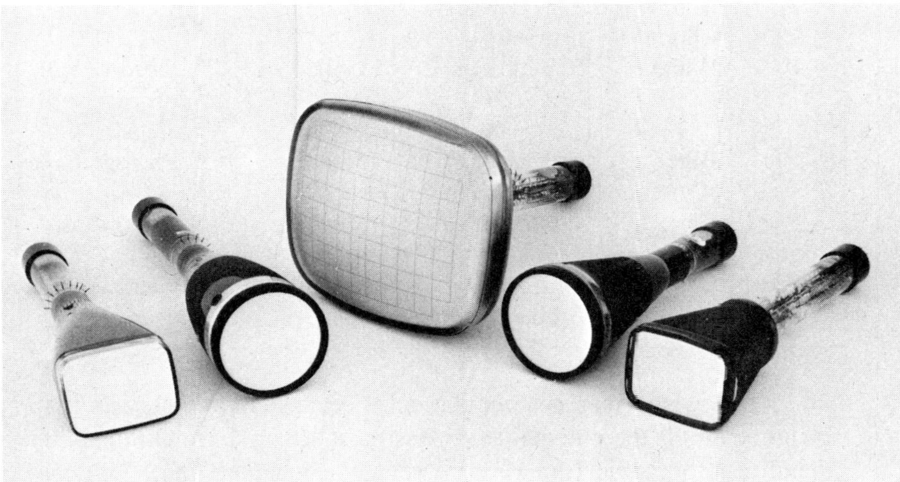

Figure 2.2 CRT screens. Note how there is more efficient utilization of the screen area with a rectangular shape. The two CRT's on the left have neck pins (*Courtesy Hewlett-Packard Company*).

phor is made of zinc orthosilicate with manganese as the activator. Wherever electrons strike it with sufficient force it becomes *luminescent*.

Luminescence is a cold light caused by the absorption of radiant or corpuscular energy, so it is different from incandescence, which is light from a heated substance. Luminescence takes place in two stages: *fluorescence* and *phosphorescence*. The first happens at the moment of impact of the electrons; the second is an afterglow that persists for longer or shorter periods in different phosphors.

Over fifty phosphors have been developed for various uses, but only the following are used in scope CRT's:

EIA Number	Color of Fluorescence	Color of Phosphorescence	Persistence of Phosphorescence	Application
P1	Yellowish green	Yellowish-green	Medium	General use
P2	Yellowish green	Yellowish-green	Medium	General use
P7	White	Yellowish-green	Long	Observation of low-speed signals
P11	Blue	Blue	Medium short	Photography
P31	Green	Green	Medium short	General use
P35	Yellowish green	Greenish-blue	Medium short	General use

Yellowish-green is a popular color because its wavelength (550 nanometers) is the one that the eye is most sensitive to. On the other hand, photographic film is more sensitive to blue. The P31 phosphor has replaced the P1 in many "late-model" CRT's because it is about twice as bright and much more burn-resistant.

Resistance to burning is important. Only 10 percent of the electron energy is converted to light. The other 90 percent turns to heat, which must be dissipated, or it will build up and cook the phosphor. This can happen if electrons continuously strike the screen at the same spot, especially in excessive numbers. The phosphor's ability to luminesce is reduced, or even destroyed altogether. To prevent this, never set the intensity control higher than necessary, and don't let the electron beam remain stationary on one spot if it can be avoided.

Burn damage arises most often because the scope is used in areas where bright lighting makes it hard to view the display. At one time it was necessary to place a hood over the screen to shield it from glare and reflections. This is seldom necessary with the P31 phosphor, especially since screens are aluminized nowadays, and given a dull surface.

Aluminizing is a process where a very, very thin film of the metal is deposited on the inner side of the phosphor. This is transparent to electrons but not to light. The electron beam passes through it and excites the phosphor, but instead of half the light being wasted by shining into the interior of the CRT it is reflected by the aluminum "mirror" and emerges from the front. This increases contrast as well as brightness, because the light is not reflected back to the screen from the inside of the tube. The dark areas therefore appear darker.

Improvement in brightness and contrast by aluminizing, combined with more efficient phosphors, can be enhanced further by etching the outside glass surface of the screen to decrease reflections. The outer surface is now usually covered with an *implosion panel* to protect it from being accidentally shattered. The panel may be either glass or plastic. A glass panel is acid-etched or coated with an antireflection coating. A plastic panel is mechanically brushed or molded to diffuse light. Another method, the most expensive, is laminating, in which a colored or polarizing film is sandwiched between the screen and the panel, which is then etched or coated.

You can also improve the contrast by using a *light filter*. This is a colored plastic plate that mounts in front of the screen. The color of the filter is matched to the color of the display. This darkens the background so the display stands out better. The ultimate in light filters is the mesh filter, which is a fine metal screen with a matte black surface. This completely eliminates reflections, but reduces light transmission from the screen as well. However, it may be the only answer in extremely bright surroundings.

We can't get much information of practical value from the display on the screen unless we have some means of measuring it. As in most cases the dimensions to be measured are vertical and horizontal a *graticule* or grid of horizontal and vertical lines is used. In many CRT's this consists of fine black lines on the inner surface of the screen in the same plane as the phosphor. This avoids *parallax error,* which is the uncertainty in measurement caused by the space between the graticule, when it is front of the glass screen, and display. In this case the apparent position of a point on the display with reference to the graticule depends upon the position of your eye, and changes if you move. If the scope you are using has an external graticule, you can avoid parallax error only by keeping your viewing angle constant. One way to do this is to use *two* graticule plates superimposed, keeping the lines in register when taking a reading. Or, if you have only one graticule plate, you can scribe a few lines on the other side of it to do the same thing.

At one time the lines on graticules were always spaced at exactly one-centimeter or one-inch intervals. You would adjust the amplification of the vertical amplifier to give a scale of "one volt per centimeter," for example, so that a waveform with a height on the screen of two centimeters would have an amplitude of two volts. Today, on an increasing number of scopes the graticule lines are at some other interval, so the scale is more often given as "one volt (millivolt, etc.) per division."

There are also polar or circular graticules for use in examining television color signals, and other graticules for special uses, but the rectangular type is the standard one. All types of graticule may be provided with edge lighting to make them more visible. This is of great help in making the lines stand out for photography.

The ratio between the dimensions of the display and the electrical function it portrays depends upon the setting of two controls on the scope's front panel. One of these determines the amount of amplification to be given to its vertical component. The settings of this control are given in volts or millivolts per graticule division, as mentioned above. If the signal has a real amplitude of one volt, and we set the control to "1 volt/division," the display will fit exactly between two horizontal graticule lines, assuming we position it to do so. If we reset the control to ".5 volt/division," the displayed waveform will double in height and occupy the space between the upper and lower of three horizontal lines. On the other hand, if we reset the control to "2 volts/division," the display will be only half the height of one vertical division, because it would take a two-volt signal to occupy a whole division at this setting.

Suppose our scope is displaying a waveform like that in Figure 2.3, and we want to measure its amplitude. The first thing we do is adjust the "volts/division" control until we get the largest display

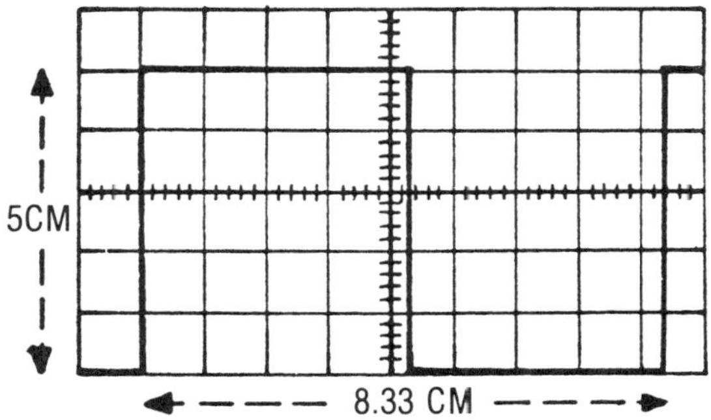

Figure 2.3 How to measure the amplitude of a waveform. (a) The amplitude of this wave form is 5cm or divisions in the vertical direction. If the VOLTS/DIV control is set to 2, the peak-to-peak voltage is $5 \times 2 = 10$ V.
(b) The horizontal distance from the beginning of the first cycle to the beginning of the next is 8.33 cm or divisions. If the SEC/DIV control is set to 2m the time interval is $8.33 \times 2 = 16.7$ ms, or 0.0167 s. Since this is the time of one cycle the frequency is given by its reciprocal (60 Hz).

that will fit between the top and bottom lines of the graticule. If this control is the kind that moves in steps (most of them are), it will have an auxiliary knob marked "variable." This must be rotated completely to the position marked "calibrated." (We'll go into the reason for this, and also what you must do if your control is of the continuously-variable type, when we discuss the vertical amplifier in Chapter 3.)

By getting the largest display possible we increase the accuracy of our measurements. We now position the display so the lower edge of it is on the bottom line of the graticule, and count the divisions and tenths of a division between the bottom line and the upper edge of the display, as in Figure 2.3. This measurement is then multiplied by the setting of the volts/division control, and the answer is the peak-to-peak voltage.

The horizontal dimension is similar, except that we are now measuring time. The sweep speed is given in units of time per graticule division: seconds, milliseconds or microseconds. A sweep speed of one millisecond per division, for instance, means that the electron beam takes one millisecond to sweep over each graticule division. If there are ten of these, it will take 0.01 second to complete each sweep across the screen. Another way of putting it would be to say it crosses the screen a hundred times each second. In fact, some scopes do have this control marked in frequency instead of time, but this will only be found in *synchronized* sweeps rather than in *triggered* sweeps. (We'll go into these terms fully in Chapters 4 and 5.)

You use the same procedure to measure the time interval between two "events" or points on the waveform as you did to measure amplitude. First, rotate the time/division control to place the points as far apart as possible (making sure the "variable" control is in its calibrated position). Then measure the distance between the two points in graticule divisions and tenths of a division. Multiply this by the setting of the time/division control. If the interval measured is from the beginning of one cycle to the beginning of the next (one complete cycle), the reciprocal of the time will give the frequency of the signal. For instance, the horizontal measurement of one complete cycle in Figure 2.3 is 8.33 divisions. The time/division control is set at 2 milliseconds per division. The time interval is therefore 8.33 x 2 = 16.7 milliseconds, or 0.0167 second. Since this is the time of one cycle the frequency is given by 1/0.0167 = 60 hertz.

From the foregoing example you can see how essential the graticule is to get the most out of your scope's screen. It takes the place of the squares on a piece of graph paper on which, in this case, you have "plotted" a graph of voltage against time. On a graph the vertical dimension is called the Y axis, and some scopes use Y to designate the vertical amplifier and its input connector. The horizontal dimension is called the X axis, so the horizontal amplifier may be called the X amplifier and its external input connector the X input.

But now a catch. If the Y axis is the vertical dimension and the X axis the horizontal direction, whatever is the "Z axis"? Obviously there is no place for a third dimension on a two-dimensional screen! However, you will find that many scopes provide for what is called *Z-axis modulation*. This is a means of connecting an external signal to the cathode or control grid of the CRT in order to modulate the electron flow as in an amplifier tube. This causes the number of electrons in the beam to fluctuate so that the beam strikes the phosphor with varying intensity. As a result, the display is brightened at those points where the intensity increases, or blanked out where the beam is cut off. Z-axis modulation has some uses in blanking out undesired portions of the display or marking points on it, as we shall see later.

The *cathode* is the element in the electron gun (see Figure 2.4) which emits the electrons that the gun forms into a beam. It is a small nickel cup, with its base facing toward the screen end of the CRT. This base is coated with barium-calcium-strontium oxide. Inside the cup is a twisted tungsten wire insulated with aluminum oxide. This wire is the heater for the cathode, and in most cases is supplied with a current of 6.3 volts a.c. or d.c. It heats the cathode to 775°C, at which temperature the oxide coating emits electrons copiously.

Surrounding the cathode is a larger cup, also of nickel. The base of this has an aperture a little smaller than the cathode. The bases of the two cups are several thousandths of an inch from each other. This electrode is the *control grid,* and has a negative potential relative to the cathode. The potential may be varied by means of the *intensity control.* The more negative it becomes, the more it repels the electrons emitted by the cathode, reducing the number that can escape through its aperture.

Beyond the control grid is the *accelerating grid,* which is a nickel cup also. However, its base is toward the control grid, and separated from it by a gap of 0.02 to 0.25 inch. The negative

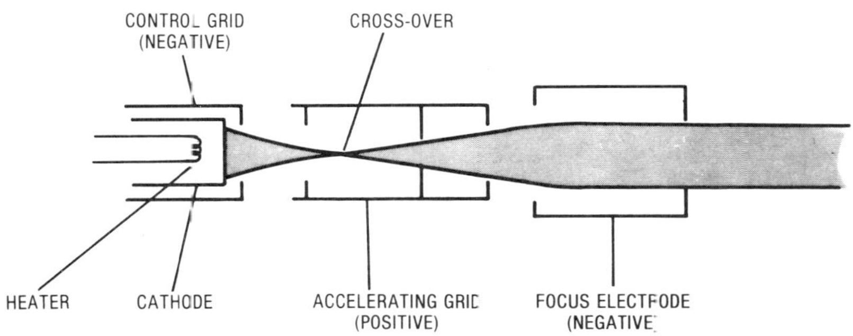

Figure 2.4 How the main elements of the electron gun form the electron beam.

potential of the control grid has a pinching effect on the electrons emerging through its aperture. The accelerating grid has a positive potential that reaches into the control grid's aperture and draws electrons through it. The resultant electrostatic field made up of the combined fields of both grids causes the electrons to converge in the gap, cross over, and diverge within the accelerating grid. (In one type of CRT there is a *laminar-flow* gun in which the electrons converge and diverge without actually crossing over.) The divergence is reduced by a second aperture that works like a stop in a camera, allowing only the electrons closest to the axis of the beam to pass and trapping the others. This reduces the intensity of the beam, but is necessary to avoid *spherical aberration,* a term used of a lens system where all the rays of light do not come to the same focus.

After the accelerating grid, the electron beam passes through the *focusing system* or ''electron lens.'' Negative potentials here force the electrons into convergent paths that give the beam the shape of a long narrow cone terminating on the screen. The *focus control* adjusts the focusing potential to make the area of impact as small as possible. The focusing system usually consists of several electrodes. In addition to the focusing electrode there is the *astigmatism* electrode that pulls the beam into a more circular cross-section. Most scopes have an *astigmatism control,* either on the front panel or as an internal adjustment. The electrode has a positive potential similar to that on the accelerating grid, but the electrons are now traveling so fast that it does not have anything like as diverging an effect. Some CRT's also have a *geometry* adjustment that varies the potential on a third element to correct for distortion similar to what is called a ''barrel'' or ''pincushion'' effect on a TV picture tube, where vertical and horizontal lines near the edges of the screen are bowed outward or inward.

Once we have got the electron beam formed to give a properly focused spot on the screen, we need a means of manipulating it to trace out the image of whatever it is we want to display. This is usually the waveform of an alternating voltage such as a sine wave, which repeats its alternations at a steady rate. The components of this waveform, voltage and time, are shown like a graph on paper, as already mentioned.

The scope ''draws'' a replica of this graph by deflecting the electron beam vertically a distance proportional to the voltage, and horizontally at a rate which is in step with the sine wave's frequency. The spot on the CRT now has two motions: a steady movement from left to right, and an up-and-down movement that, combined, result in the representation of the waveform.

If the frequency is low enough, we see the spot traveling steadily across the screen, rising and falling as it goes. On reaching the right-hand edge it disappears and reappears immediately at the left to begin another sweep from left to right. At higher frequencies the spot has to travel faster to maintain synchronization with the sine wave, and the spot becomes a line. It is now sweeping across the screen hundreds, even thousands, of times a second, and the glow of the phosphor has hardly begun to decay before the electron beam excites it again.

These movements are produced by varying the voltages on two pairs of *deflection plates* within the CRT. The electron beam passes between these plates after leaving the electron gun. It passes first between the *horizontal deflection plates.** At the beginning of each sweep there is a high positive voltage on the left-hand plate and a much lower or even negative voltage on the right-hand plate. Since electrons are negative they are attracted toward the left-hand plate, but because of their high speed they do not actually strike it, but are deflected so that they impinge on the left-hand side of the screen. But the voltage on the left-hand plate is decreasing at a steady rate while that on the right-hand plate is increasing, so the electron beam is attracted less and less to the left-hand plate and more and more to the other. At the end of the sweep the voltages on the plates revert to their original values and the process repeats itself (except in the case of "one-shot sweeps," of which more hereafter).

After passing between the horizontal deflection plates the electron beam passes between the *vertical deflection plates.* The principle of operation is the same as for the horizontal plates, except that the plate potentials vary in proportion to the amplitude of the input signal. In the absence of an input signal the beam would still sweep across the screen (if set to do so), tracing a straight line at a level determined by the setting of the *vertical position control.* The appli-

---

*In scopes designed for very high frequencies, where higher anode voltages are often used, the vertical deflection plates may be positioned first (nearer the cathode) to offset the "stiffer" beam.

cation of a postive d.c. voltage would normally move it upward by increasing the positive potential on the upper plate. A negative input voltage would have the opposite effect, by repelling the beam toward the other plate. As we have seen, the fluctuating potential of an a.c. input causes the beam to move up and down in conformity with it.

The maximum angle of deflection of the beam should not exceed 15 degrees if distortion is to be avoided, which explains why scope CRT's are long and narrow compared with picture tubes. The actual angle depends upon the velocity of the electrons and the plate potentials. High-speed electrons produce a brighter display, but require higher deflection voltages. Since these would require more powerful amplifiers, which would add to the cost of the instrument, scope manufacturers looked for a way of increasing brightness without increasing deflection potentials. They found it in *post-deflection acceleration* (PDA).

With PDA the electron beam is only moderately accelerated up to the deflection plates, so it is reasonably flexible as it passes between them. As in a picture tube, the inside of all CRT's originally had a coating of aquadag graphite. In CRT's using PDA, however, this coating is in the form of a narrow ribbon of resistive material arranged in the form of a helical spiral, extending from a coating in the neck close to the deflection plates all the way to the screen. The PDA voltage is applied across this spiral, which acts like a continuous voltage divider to set up a potential gradient extending practically the entire distance from the gun to the screen. The electrons are progressively accelerated by the rising strength of its electric field until they strike the phosphor.

The beam electrons are called *primary electrons*. Where they strike the screen, energy is transferred from them to the electrons in the phosphor. Some of these are raised to energy levels that cause them to diffuse to the surface and escape into the vacuum in the tube. These are called *secondary electrons,* and the process is called *secondary emission*. The secondary electrons are collected by attraction to the high positive potential on the coating and also the aluminum film on the phosphor. From here they return via the power supply to the cathode, thus completing the circuit.

Secondary emission is the secret of the *storage tube*. This CRT has an electron gun just like the one in a conventional CRT. In the storage CRT this gun is called the *writing gun*. There are also two other guns called *flood guns*. These are installed on opposite sides of the neck of the tube, but have no effect on the writing gun beam. The flood guns emit two cones of electrons which spread out evenly over the *target,* which is the name for the screen in this tube (see Figure 2.5).

When the writing gun is not operating, only the flood-gun electrons land on the target. The potential difference between the flood-gun cathodes and the *storage target back plate* (STB), or collector, is quite low, about 150 volts, so these electrons do not have enough energy to make the phosphor glow. For the same reason, secondary emission is low. In fact, more electrons land on the target than leave it. As a result the target acquires a negative charge.

As this negative charge builds up it forces the flood-gun electrons to slow down more and more as they approach the target. Primary electron current (beam current) gets less, consequently secondary emission decreases. In the end the primary and secondary electron currents reach a stable, balanced condition called the *rest potential,* which is much lower than the STB voltage.

When the writing gun operates, its beam of high-energy electrons causes the phosphor to glow where it strikes it, and to emit large numbers of secondary electrons. This part of the target becomes depleted of electrons, which means that the visible display has a positive charge. The potential difference between the *written area* of the target and the flood-gun cathodes is therefore much greater, and the flood-gun electrons landing on it do so with a lot more energy.

The effect of this is that even after the writing gun stops, the flood-gun electrons continue to be accelerated by the positive charge on the written area and strike it with sufficient energy to keep it glowing. As the emission of secondary electrons now exceeds the supply of primary electrons the positive charge is maintained. In this way a second stable state has been attained.

Since any point on the screen of the storage tube will always be in one or other of these two stable states it is called a *bistable storage tube,* and as it produces a visible display it is also called a *direct-view CRT*. The bistable CRT stores all events with the same level of brightness, but there is another type of tube called a *halftone tube* that can display the stored signal at different levels of brightness. This is done by variations in the beam current. There is also an *electrical-readout tube* that does not have a visible output. This is used with computer circuitry, so we shall not consider it in discussing scopes.

The stored image will remain on the screen indefinitely unless the scope is turned off, or until the screen is *erased.* To do this a push-button control marked ''erase'' is used. When you depress it a voltage pulse is applied to the backplate. This raises the target to the upper stable state all over. It is followed by a negative pulse that takes the target all the way down to well below the rest potential. The backplate voltage then comes up slowly to its normal value, leaving the whole target at the lower stable state.

It is common practice for storage tubes to have a *split screen.* The target is made in two sections isolated from each other. Each half has its own ''store,'' ''erase'' and ''off'' push buttons. Depressing the ''store'' push button applies the operating voltage to the backplate for storage operation; depressing the ''off'' button removes it. With this arrangement you can store a display on one half of the screen while you write another on the other half, either for storage or conventionally. This capability has many uses. For instance, when used in speech therapy a normal speech pattern is recorded on one half of the screen and the patient's attempt to copy it is displayed on the other half for comparison.

By adding a mesh dielectric screen just behind the phosphor screen we can store a display on it instead of directly on the screen. The display is stored in the form of positive and negative voltages, so is invisible itself. However, flood-gun electrons are repelled by the negatively-charged areas and accelerated by the positively charged areas. The accelerated electrons pass through the mesh to strike the screen and reproduce the image in visible form. This technique is used to capture and store high-speed signals that travel too fast to raise the target to the upper level if applied directly to it, and is called *transmission storage.*

Halftone storage, mentioned above, is also called *variable persistence.* It allows you to set the length of time that the image will stay on the screen. If you are trying to study a signal that is accompanied by much noise or jitter, variable persistence enables you to set the duration of the image so that only the repetitive parts of the signal remain on the screen. The transients fade too fast to be seen.

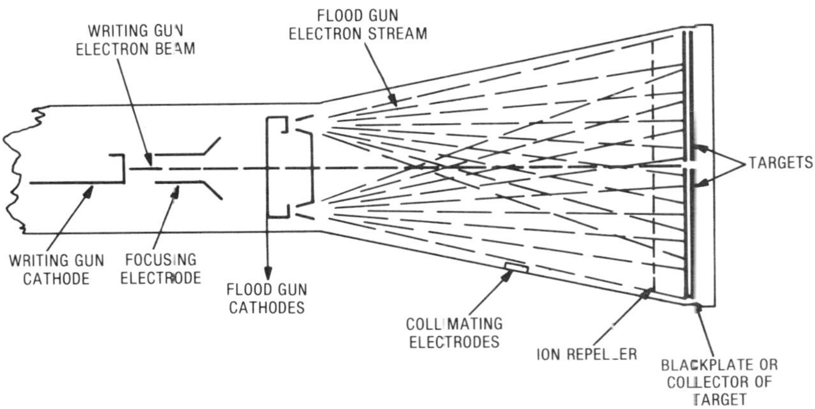

Figure 2.5 In the storage CRT there are two "flood" guns in addition to the conventional "writing" gun. The screen, or "target," is usually split into an upper and lower section. So that half the screen can be used for storage and half for a normal display, if desired.

Two other elements shown in Figure 2.5 are the *ion repeller* and the *collimating electrodes*. The ion repeller is a metal screen with a fine mesh mounted about five millimeters in front of the target. This screen has a voltage somewhat higher than the backplate, around 175 V. The mesh is fine enough to create a uniform field that repels positive ions, but is coarse enough to have negligible effect on electrons passing through it. Positive ions are cathode-material atoms that have lost an electron (hence the positive charge), and have acquired sufficient thermionic energy to escape. Owing to their much greater mass they are not easily deflected, and generally hit the center of the target if some means of stopping them is not provided. You'll recall that the same thing would happen in a TV picture tube if it didn't have an ion trap. They would do considerable damage to the phosphor, producing a brown spot called an *ion burn* if it were not for the ion repeller.

A second task performed by the ion repeller is the elimination of *trace shadowing*. This is a condition caused by the positive charge on the written area. This charge not only accelerates the flood-gun electrons that would land on the written area anyway, but also pulls in those electrons that would normally land just outside the area. In average operation there is a certain amount of background illumination in the unwritten areas of the target (caused by flood-gun electrons), so this pulling-in of adjacent electrons produces an undesirable dark border on each side of the trace. The ion repeller, by creating a homogeneous field in the area between it and the target, overcomes this effect.

The collimating electrodes are necessary for uniform distribution of the flood-gun electrons over the target. With a potential of about +150 V compared to the flood-gun cathodes, they cause the electrons to spread out in two cones of even density right to the edge of the target.

On most storage scopes you will also find CRT controls called "enhance," "integrate" and "auto erase." The "enhance" control is used with "single-shot" sweeps, where only one excursion of the writing beam across the screen is not enough to raise the written area of the target to the upper stable level. In the absence of further sweeps the target will go back to the rest potential. The "enhance" control sends a pulse to the STB that raises the target potential sufficiently so that the written area is lifted to the upper stable level. It then stays at this level when the pulse is removed, while the unwritten area returns to the rest potential.

The "integrate" control is for repetitive traces that are too fast to raise the target to the upper stable level in one pass. In between sweeps the flood-gun electrons neutralize the small positive charge on the written area. By depressing the "integrate" push-button the flood-gun electrons are cut off. The writing-gun beam continues to sweep over the same area, but now there are no flood-gun electrons to replace those it is ripping out of the phosphor. The written area becomes more positive with each sweep. When the "integrate" push button is released the written area potential has risen to the upper stable level and the display is stored.

"Auto erase" allows the display to be automatically erased after a set interval, which may be up to ten seconds.

The phosphors used in storage tubes are the same that are standard in conventional CRT's, P1 or P31. The granules of phosphor are deposited in a thin layer. This allows the light to pass through, and the particles are not close enough to short together (if they did the image would disappear). The STB is a transparent conductive coating on the inside of the face plate, between it and the phosphor.

There is one further CRT control found on many scopes both conventional and storage. We have left it until last because it does not operate directly on the CRT although it is usually placed among the CRT controls already discussed. This control is generally in the form of a push button labeled "beam finder." It sometimes happens that the display is not present on the screen although the scope is operating. Increasing the intensity or juggling with the other controls does not always get it on the screen. The "beam finder" reduces the drive to the deflection plates so that the display appears in miniature form on the screen, and we can see in what direction it lies if it is outside the screen area when it is full size. Then it is just a matter of adjusting the positioning controls in the right directions to bring it on. Sometimes the display is on the screen, but not visible for another reason, such as an untriggered sweep. The "beam finder" will reveal it, and if we do not see it after repositioning it we can check other controls that might be responsible. We shall cover this later.

Replacing the CRT

Should it be necessary to replace the CRT in your scope, you can do it yourself, provided you use common sense and are careful. There is a high vacuum within the glass envelope, so if you break the latter the air outside rushes in with great force. This is called an *implosion,* to distinguish it from an explosion, in which the force is directed outwards. It is just about as dangerous either way, since it doesn't matter which way the piece of glass that hits you in the eye was going initially! So obviously you should wear safety glasses, and avoid striking the CRT on any object that might cause it to crack and shatter.

A replacement tube of the same type as the old one can be ordered from the scope manufacturer. In many cases you can also get one from an electronics supply house, but if the manufacturer makes his own CRT's you will have to get it from him, since only his are tailor-made for the instrument.

To remove the old CRT, you begin by taking off the bezel, which is the frame or rim around the screen. Usually this is secured at each of its four corners by a nut or screw. Save these, as you will need them later when you put it back. After removing the bezel, you

TROUBLESHOOTING CHART FOR CRT WRITING GUN CIRCUIT (If this is a storage tube, set storage switches to off to protect target.)	TROUBLE SYMPTOMS							
	No display	Low or high intensity	Poor focus or astigmatism	Blooming	Other abnormal symptoms	Linearity poor	Display tilted	Screen burnt
Intensity control set improperly	X	X						X
Display positioned off screen	X							
Unblanking circuit inoperative	X							
Line voltage missing or too low	X	X	X		X			
Low-voltage power supply defect	X	X			X			
High-voltage rectifier(s) defective	X	X			X			
High-voltage divider defective		X						
CRT cathode voltage incorrect	X	X			X			
CRT grid voltage incorrect	X	X			X			
Astigmatism control set improperly			X					
Focus control set improperly			X					
Anode voltage incorrect		X		X	X			
Gassy CRT				X				
High-voltage filter capacitors		X			X			
Geometry control misadjusted						X		
Tube turned, clamp loose							X	

Chart 2.1

TROUBLESHOOTING CHART FOR CRT STORAGE CIRCUIT (Set storage switches to "store" with enhancement mode "off.")	TROUBLE SYMPTOMS							
	Screen does not store	Screen does not erase	Poor contrast	Screen does not store single sweep display	Background brightens after storage	Enhance mode inoperative	Integrate mode inoperative	Other abnormal symptoms
CRT not seating properly	X	X						X
Upper and lower storage voltages incorrect	X		X		X	X		X
Low-voltage power supply incorrect	X							X
Flood gun(s) inoperative	X							
Storage circuit voltages incorrect	X	X	X	X	X	X		X
Enhance mode control set improperly			X		X			
Sweep too fast				X				
Storage screen weak (replace CRT)				X				
Intensity control set improperly				X				
Integrate circuit voltage incorrect							X	

Chart 2.2

also remove the plastic graticule if it is not scribed in the CRT screen itself.

Next, you disconnect the four leads going to the deflection plates, if this is the type of CRT in which the deflection plate pins are located on the neck of the tube just before the beginning of the flare. In other CRT's, however, the deflection plates are connected to pins on the base of the tube. Pull the lead connectors straight off, noting which goes to which pin (they are usually color-coded).

Disconnect the CRT base socket by pulling it gently straight back. In some scopes a plastic handle is provided, as part of the socket, to make it easier, as there may not be much room for your hand. In CRT's with a separate high-voltage lead, disconnect this also.

There is usually a base clamp, which must be loosened next. It may require a screw driver or an Allen wrench.

The CRT should now be ready to come out. Grasp its face with your right hand, and push it carefully toward the front of the scope with your left hand. Be careful not to bend the base pins, or the deflection plate pins on the side, if it has them. Remove the CRT through the front of the scope.

Carefully insert the new tube into the shield and the tube base clamp. See that the deflection plate pins on the side, if it has them, are correctly positioned. If there are wall bands or other contacts on the wide end of the tube, make sure they are also aligned to make contact with the metal contact fingers in the tube shield. This only applies to scopes that have a tube shield, of course. Reconnect the high-voltage lead, if there is one.

Reconnect the CRT socket to the tube base, and the deflection plate leads to the correct pins. Install the CRT graticule (if it has a separate one) and bezel. Then turn on the scope power and obtain a free-running trace. Rotate the CRT as necessary to align the trace horizontally. Then tighten the clamp (but do not over-tighten it).

You will then have to recalibrate all circuits except the low-voltage power supplies.

Store the old tube in the box the new one came in, and arrange for a safe disposal of it. Sometimes it is required as a trade-in, but in any event be careful with it!

Multi-beam CRT's

Before leaving the subject of CRT's let us take up some special types called multi-beam CRT's. These are made in three different types: split-beam, dual-beam and dual-gun. In practice, any of the three may be called dual beam.

A split-beam CRT has a single electron source up through the second anode, at which point the beam is split. Each beam is then sent through independent vertical deflection systems and through a common horizontal deflection system.

Some of the earlier split-beam CRT's were single-ended: one plate of the vertical deflection system was common to both beams. Each beam had independent deflections, but shared the common center plate. Newer split-beam designs are double-ended with two fully separate sets of vertical deflection plates.

Split beam represents a giant stride forward, compared to dual trace (see Chapter 3) because time sharing is no longer involved. The traces are displayed 100 percent of the time. Split beam also overcomes the other major disadvantages of dual trace: there is no signal loss or extensive light loss due to time sharing; a single horizontal position control maintains time registration on all displays; and the single horizontal amplifier minimizes the time adjustments for the split beam.

Among split-beam limitations is lower writing speed due to only one electron source being available for both beams. Also, the single electron source ties both beams to a common intensity control. If Z-axis modulation is needed for a given application, this single electron source prevents independent modulation.

Dual-Beam CRT's use two electron sources, two independent vertical deflection systems and a common set of horizontal deflection plates. They overcome the same dual-trace disadvantages as do split-beam scopes and, additionally, overcome the electron source limitations of split beam. Since they use a single horizontal amplifier, a single time base and a single trigger circuit, dual-beam units are less costly than later developed dual-gun instruments.

The third dual-beam CRT and the one representing the most advanced technology, is the *dual-gun CRT*. This CRT has two totally independent systems: two electron sources, two vertical deflection systems and two horizontal deflection systems. Only the display screen is shared by both beam systems.

Undoubtedly, the principal unique advantage of a dual-gun scope is its ability to display one signal—or a single event—in real time at two different sweep speeds. Almost as important is the dual gun's ability to trigger each sweep separately. Thus, displays of two signals, unrelated in time, are stable.

In practice, all dual-gun scopes have two vertical amplifier sections, two horizontal amplifier sections and two horizontal time bases. Selective cross coupling is also available on newer dual-gun scopes so that either or both beams can be driven by any combination of vertical or horizontal systems.

A primary disadvantage of the dual-gun scope is that each beam has a separate horizontal position control. When making time-related measurements, you have to register each beam with the other, using two separate controls. This can be an inconvenience when compared to scopes with dual-beam or split-beam CRT's.

Power Supplies

There are two different power supply requirements in a scope:

Low-Voltage Power Supply
High-Voltage Power Supply

Low-Voltage Power Supply

The low-voltage power supply converts the a.c. power-line supply into one or more d.c. voltages suitable for the d.c. power supplies required by tubes or transistors. This circuit must do three things:

Rectify the alternating sine wave so it becomes pulsating d.c.

Smooth out the pulsations and regulate the d.c. voltage levels so they remain constant under varying loads and in spite of power-line fluctuations.

Step voltage up or down as necessary for various supplies.

The last-named requirement is provided for by the use of a power transformer with taps on its secondary winding from which various a.c. voltages are obtained. These voltages will be of the proper values to give the d.c. levels required after rectification, filtering and regulation.

Rectifiers are diodes, and may be vacuum tube or solid state. The solid state types include selenium, germanium and silicon diodes, with silicon being the preferred type. A single rectifier can be used in a half-wave rectifier circuit, but a full-wave rectifier is preferred since it uses both the positive-going and negative-going portions, and so requires less filtering.

Filtering in simple power supplies is most often accomplished using an RC filter. The time constant of the circuit is chosen to limit the rapidity of voltage variations at the output. However, it does not do as complete a filtering job as we'd like. Furthermore, the voltage drop across the series resistor subtracts from the output voltage. Substituting an inductor for the resistor avoids this voltage drop, so the preferred type of filter is the pi filter, with two capacitors and an inductor (Figure 2.6).

An even more sophisticated approach uses negative feedback, as in the circuit in Figure 2.7. This circuit combines the task of eliminating the ripple with that of maintaining a constant voltage level in spite of changes in the load or variations in the line voltage.

Instead of the inductor of Figure 2.6 this circuit has a tube (V1) called the series-regulator tube. It operates like a potentiometer, since the voltage on its grid controls its internal resistance. The grid voltage is supplied by the plate voltage of V2.

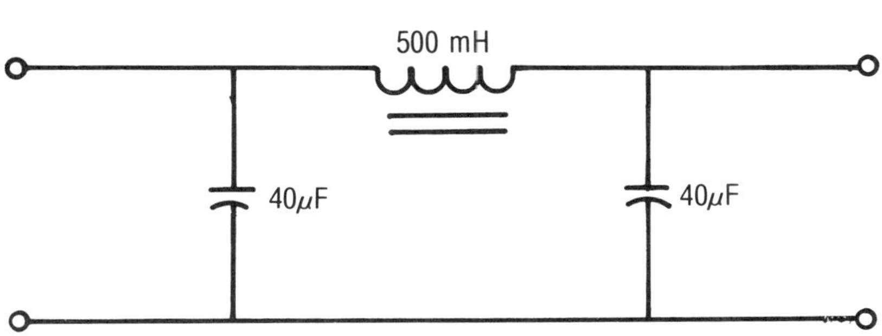

Figure 2.6 The π filter smooths the input from a full-wave rectifier so that the output is virtually pure d.c.

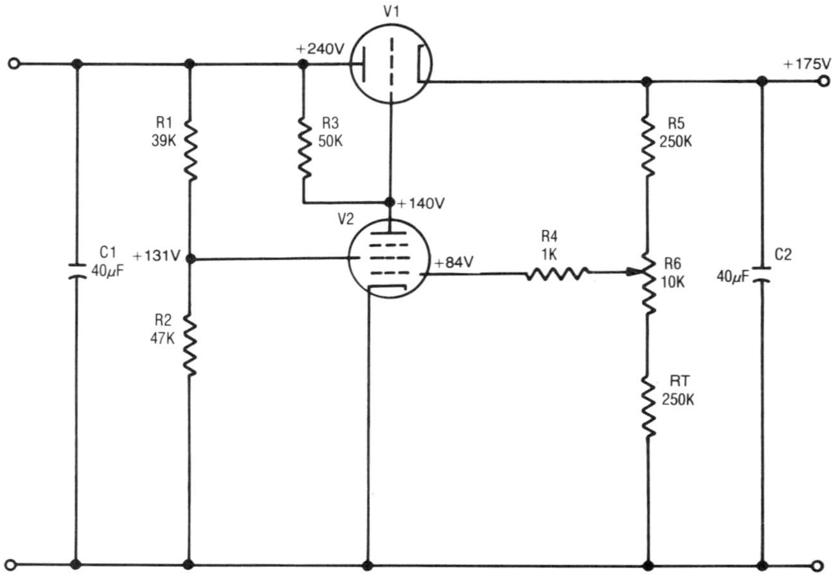

Figure 2.7 Filtering by negative feedback.

V2 derives its grid voltage from the voltage divider R5, R6 and R7. A change in the voltage at the cathode of V1 will cause the voltages dropped across this divider to change also. An increase in voltage will result in V2's grid going more positive. This increases the plate current in V2 and in R3. The increase of current in R3 gives a greater voltage drop across this resistor, so that the voltage at V2's plate, and on V1's grid, is lowered. This increases the internal resistance of V1, so more voltage is dropped across it, restoring its cathode voltage to the proper value. A decrease in V1's cathode voltage has the opposite effect, of course.

This feedback circuit not only maintains the output voltage at the proper value, but cancels the ripple signal on the plate of V1. Any ripple component appearing on the cathode of V1 is amplified by V2 and appears on V1's grid with the opposite phase, to oppose the ripple on V1's plate and reduce the amplitude of ripple current to a very low value.

R6 is adjustable to allow the output voltage to be set to the exact level desired. It is not provided in all power supplies. Another optional feature may be a neon glow tube in the cathode circuit of

V2 to clamp the cathode voltage to a fixed value. Similar circuits are also found in transistorized equipment.

High-Voltage Power Supply for Cathode-Ray Tube

CRT's require a high anode-to-cathode voltage. In many cases this may be as much as 12 kilovolts, but the current drawn is very small. Consequently, the type of power supply shown in Figure 2.8 is used.

The output of an oscillator V1 is stepped up by a high-voltage transformer T1. The secondary voltage is rectified by diodes and filtered by resistors and capacitors, and the DC output is applied to the CRT, positive going to the anode, negative to the cathode and grid. A sample is often fed back to a comparator which controls the oscillator.

In Figure 2.8, V1 and the primary of T1 form a Hartley oscillator, in which the inductor resonates with the distributed capacitance of the circuit. The supply voltage for this oscillator comes from the low-voltage supply of the oscilloscope. The frequency of oscillation will be in the order of 50 to 100 kilohertz. Filtering such a frequency is easier, and requires much smaller capacitors, which is an important consideration at high voltages.

Two voltages are supplied by this circuit. A positive voltage of 10 kilovolts is required for the CRT anode, and a negative voltage of 2 kilovolts for the cathode and grid.

The anode voltage is obtained by using a voltage tripler. However, the higher frequency makes filtering easier, so smaller values are required for C5, C6, C7 and C9.

The cathode voltage, being much lower, is provided by the half-wave rectifier circuit consisting of D1, C4 and C8. Because of the higher frequency and the very small load involved this simple filtering is sufficient.

R7, R8 and R9 form a voltage divider in which R9 is very much larger than the other resistors. The voltage sensed at the movable contact of R8 is only a small percentage of the total. This voltage is fed to the right-hand grid of the dual-triode V2.

The right-hand cathode of V2 is connected to the negative B supply (usually −150 volts). This supply is well-regulated, so the

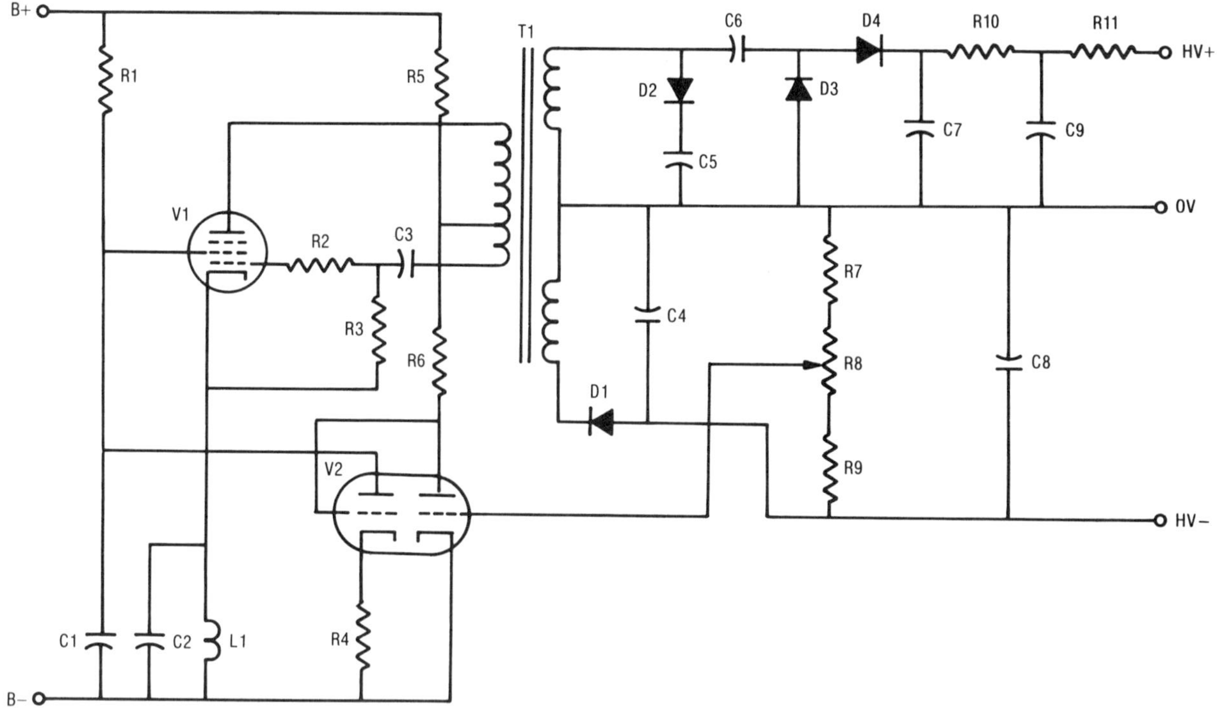

Figure 2.8 High voltage power supply for CRT

cathode voltage is constant. Consequently, any change in the grid voltage is followed by a change in the plate voltage. This *error voltage* is coupled directly to the left-hand grid, and the resultant amplified change of plate voltage appears on the screen grid of V1.

This, in turn, changes the amplitude of the Hartley-circuit oscillations. If the negative voltage on R8 decreases, the right-hand plate voltage decreases, the left-hand grid voltage decreases, the left-hand plate voltage increases and the screen-grid voltage of V1 increases. This causes the oscillations to increase in amplitude, so that the high voltage rises.

The high voltage is set to the exact potential required by adjusting R8. L1 is a choke to keep the oscillator signal from coupling into the power supply (whence it would be distributed all over the set). R1, R5 and R6 are voltage-dropping resistors. R4 is the bias resistor for the left-hand half of V2; C1 and C2 are bypass capacitors.

In some models solid state diodes may be replaced by high-voltage vacuum-tube rectifiers. V1 and V2 also may be transistors. The oscillatory circuit may be any suitable type instead of a Hartley. Part of the high voltage may go to the CRT intensity control. The anode or cathode may have independent supplies.

Finally, smaller or more economical equipment may obtain a sufficiently high voltage from a secondary on the low-voltage power supply transformer. The actual voltage requirements will be different in various models.

3

Vertical Amplifier

Many low-frequency signals could be shown perfectly well by applying them directly to the CRT vertical deflection plates, as long as their amplitudes were within the range that would give a useful display. However, this would cut down the scope's usefulness tremendously. Furthermore, the loading effect of the deflection plates' capacitance might have a considerable effect upon the performance of the circuit under test.

The vertical amplifier and probe overcome these limitations by changing the amplitude of the signal to one that is the right size for the screen, and by acting as a buffer between the CRT and the circuit. In doing so, we expect them to make a faithful reproduction of the signal as well.

If we follow the signal from the probe tip to the CRT (see Figure 3.1), we start with the probe. While simple test leads work for some low-frequency applications, a compensated probe is necessary as a general rule. It is more convenient for contacting points in the circuit, and ensures accurate measurements.

A compensated probe is illustrated in Figure 3.2(a), and its schematic is given in (b). With the variable capacitance in the probe body it is possible to match the probe time constant with that of the scope input and cable.* This probe has a 9-megohm resistor, so that it forms a voltage divider with the scope input that gives a probe attenuation factor of 10X. This is a very popular value since it applies a 10-megohm load to the circuit being probed instead of the

---

*So that $R_pC_p = R_iC_i$. As long as these time constants agree, the probe impedance Z_p and the scope input impedance Z_i maintain a constant *ratio* which is independent of frequency.

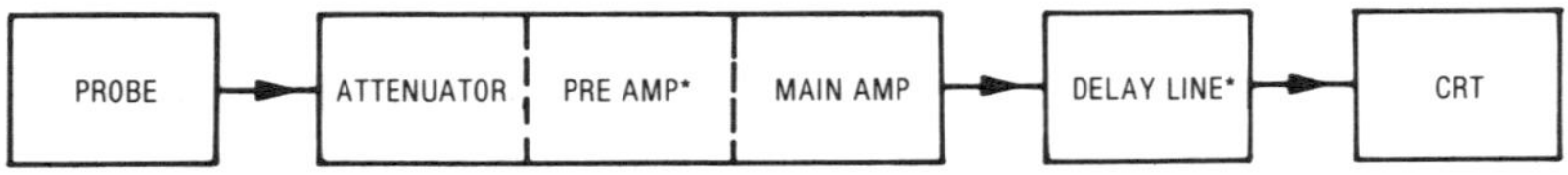

*Omitted in simpler scopes.

Figure 3.1 Signal path from probe to CRT

standard scope loading of 1 megohm, but you have to remember to multiply the amplitude measured on the CRT screen by ten to allow for the probe attenuation.

There are also probes with an X1 attenuation, but their loading (1 megohm) has a greater effect on the circuit. At the other extreme there are high-voltage probes with an attenuation factor of X1000 and an input impedance of 100 megohms that are capable of measuring 40 kilovolts.

The compensation of these probes is adjusted, in some cases, by loosening one portion of the probe and twisting until it is correctly compensated, then retightening; or by tweaking a screwdriver adjustment on the side. You apply the tip of the probe to a source of square waves, usually the scope calibrator (which will be discussed later). When properly compensated, the waveforms will have flat tops, as in Figure 3.3(a). An overcompensated probe will distort the waveform as in (b), and an undercompensated probe will give a display like that in (c). (See page 58.)

Probes of this type are called passive probes. For frequencies above 250 megahertz an active probe should be used. This contains an FET and a power supply. As you know, an FET has a very high input impedance and very low input capacitance, so the loading effect is negligible and high-frequency signals are not degraded.

Demodulator Probe

A demodulator or detector probe (Figure 3.2(c)) allows you to make accurate measurements of the modulation superimposed on very high-frequency carriers. It is most used for signal tracing in a TV video i.f. amplifier, since most service scopes do not have a bandwidth of 50-megahertz. As you can see, it contains an AM detector circuit to remove the carrier, leaving the modulation to be displayed. If the signal is not modulated, you will get nothing.

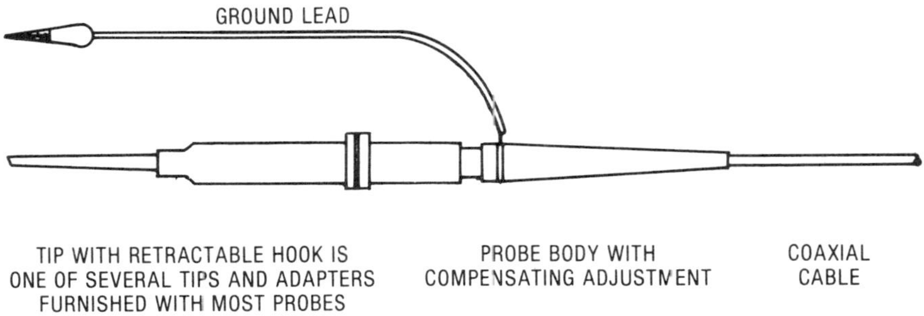

(a)

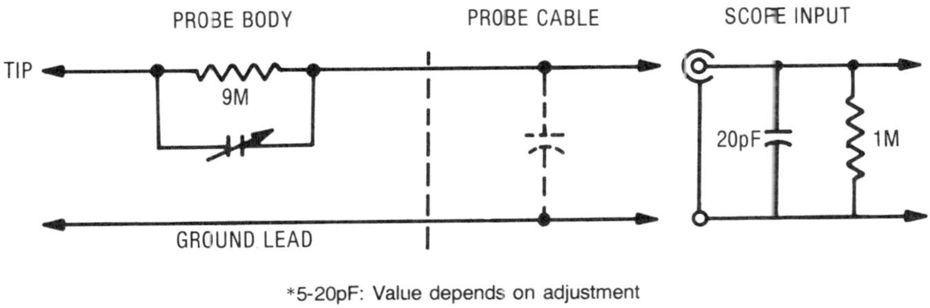

(b)

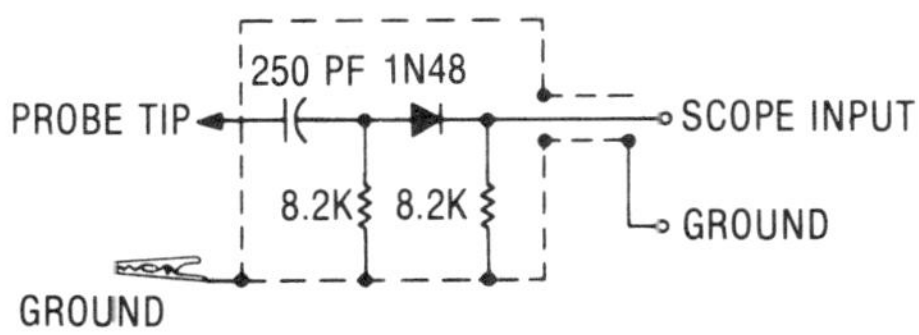

(c)

Figure 3.2 (a) Typical scope X10 compensated probe
(b) Scope input schematic
(c) Demodulation probe schematic

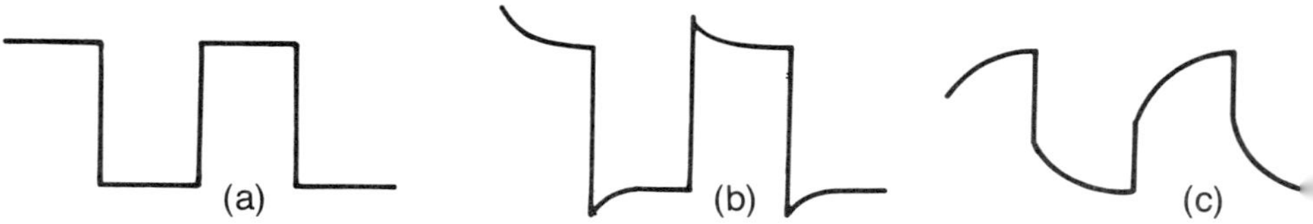

Figure 3.3 Square-wave display when probe is: (a) properly compensated; (b) over-compensated (capacitance too high); (c) under-compensated (capacitance too low).

Vertical Input

The vertical input connector on most scopes today is a BNC receptacle, and probe or other coaxial cables have a mating plug. Some may have the older type N, and for very high frequencies the GR874 (push-on). Double banana jacks were popular at one time, but they are only suitable for low frequencies, even when shielded plugs and leads are used, because of their high input capacitance.

Immediately following the input connector is a switch, usually three-way, to give you the choice of d.c. or a.c. coupling, or of grounding the input. A capacitor, typically 0.1 microfarad, blocks d.c. when the switch is in the a.c. position, otherwise it is bypassed. In the "ground" position the input is disconnected and the attenuator input is connected to ground. This is useful when you want to position the trace to give a zero reference (no signal applied) so you can have an accurate baseline from which to make amplitude measurements.

After passing through the switch, the signal comes to the attenuator, which, as mentioned above, can be set to make the signal a convenient size to fit on the screen. If the vertical amplifier is designed to amplify a 50-millivolt input signal so that it has an amplitude on the screen of one vertical division, it is said to have a *deflection factor* of 50mV/division. You will also come across the expression *deflection sensitivity,* which should be defined as so many millimeters (of screen deflection) per volt. However, there is a tendency to use these terms rather loosely and interchangeably, and you will be just as likely to come across a statement that "the sensitivity of the unit is .05 volts/cm" instead of saying that that is its deflection factor.

Be that as it may, if a 50-millivolt signal without attenuation occupies one division of the graticule, we are obviously going to have to do something about, say, a 5-volt signal, which without attenuation would in theory be 100 divisions high.

The vertical attenuator usually consists of a number of attenuation networks mounted on or nearby a switch, controlled by a knob on the front panel. A common name for this control is VOLTS/DIV. The attenuation networks divide the signal typically by factors of X1, X2, X4, X10, X20, X40, X100, X200 and X400. As the VOLTS/DIV control is turned, the attenuation networks corresponding to the values marked on the control are switched into contact with the vertical input d.c.-a.c. switch on the one hand, and the input to the input amplifier on the other. The control positions on the panel are typically marked .05, .1, .2, .5, 1, 2, 5, 10 and 20, these being the voltages that will give a one-division deflection at each position.

Figure 3.4 illustrates the first two steps of an attenuator. In the first step, corresponding to ".05" on the dial, there is no attenuation. The signal goes straight through to the grid of the amplifier tube or base of the transistor. This scope has an input impedance of one megohm shunted by 33 picofarads. This is provided by R3 and C3 (shunted by C_d, the distributed capacitance of the circuit), which are in the circuit all the time.

In the ".1" position of the switch, the X2 attenuator network is connected between the input and the input amplifier. This consists

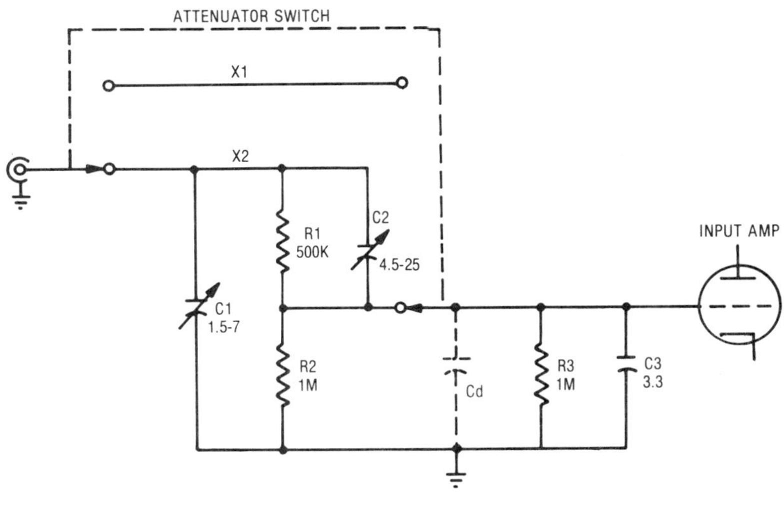

Figure 3.4 Input attenuator network (simplified)

of R1, R2, C1 and C2. R2 is one megohm, and is in parallel with R3 (also one megohm), so that their combined resistance is 500 kilohms, in series with R1 (also 500 kilohms). These two resistances, totaling one megohm, preserve the resistive part of the input impedance. At the same time, as the grid signal for the tube (or base signal for a transistor) is now taken from the middle of this voltage divider instead of from the top, its amplitude is halved.

The parallel capacitances of C3 and C_d now shunt R2. C2 is adjusted to make the RC combination of R1 and C2 equal to that of R2, R3, C_d, C3. C1 is then adjusted to bring the total input capacitance to 33 picofarads. In this way, the input impedance of the scope does not change when you switch from X1 to X2.

Attenuation networks for the other factors work in a similar manner. They are all compensated voltage dividers with adjustments to maintain the same input impedance for any range. In some cases they are "stacked": an X10 network may be followed by an X2 network, giving a total attenuation of X20.

A step attenuator always has a variable (VAR) control to enable you to adjust the amplitude of the display to any desired intermediate value between the calibrated steps. It is usually in the form of a potentiometer that alters the amplifier gain. It is important that you remember to keep this control in its calibrated (CAL) position when you are not using it, so that the attenuator steps have their proper values. (See Figures 3.6 and 3.10 for examples.)

Input Amplifier

Immediately after the attenuator comes the first amplifier stage of the input amplifier. As it is desirable that this stage shall have a high input impedance, so as not to shunt the attenuator networks, and a low output impedance, not to mention linear amplification over a wide range of input frequencies, it is almost always a cathode, emitter or source follower. A comparison of the triode, transistor and IGFET to remind you of their lead terminology is given in Figure 3.5.

The input amplifier may be built into the main body of the scope, or may be all or part of a plug-in unit. If there is also a main amplifier following, the input amplifier may be called a *preamplifier*. The degree of complexity of the vertical amplifier is

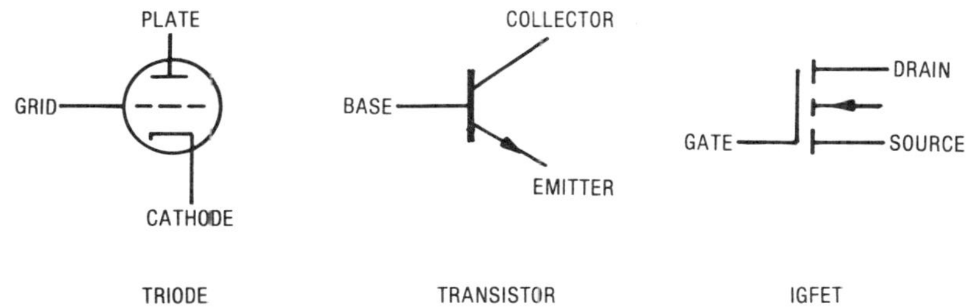

Figure 3.5 Lead terminology of triode, transistor and IGFET.

proportional to the amplification-bandwith-risetime performance required, and the type of amplifier, of which more later.

Vertical amplifiers commonly have single-ended input stages and push-pull output stages. The first stage is a cathode, emitter or source follower, as mentioned above, followed by a phase inverter. Phase inverters are also called phase splitters and paraphase amplifiers. In scopes they are all comparators with one input grounded, as in Figure 3.6, where V1B's grid is grounded. The two outputs of opposite phase then drive push-pull output stages.

An amplifier is often symbolized by the triangle of Figure 3.7(a). Its voltage gain can be estimated by dividing R_o by R_i. The common-emitter circuit shown at (b) is typical of many amplifiers. R_i comprises the emitter resistance R_E and the transistor resistance R_t. R_o is the output resistance. The voltage gain (Av) is given by $R_o/R_i = 2500/532 = 4.7$.

Delay Lines

In the vertical amplifier a sample of the input signal is diverted to the trigger and time-base circuits to synchronize the sweep. The rest of the input signal continues on through the remaining stages of the vertical amplifier. In the absence of other circuitry it may reach the deflection plates in 20 nanoseconds or so. However, the portion of the signal feeding the trigger circuits only starts a chain of events: switching, shaping, generation and amplification all have to take place before the horizontal sweep begins. This sequence may take about 100 nanoseconds. As a result the first 80 nanoseconds of the signal cannot be seen. In higher-frequency signals this is a considerable portion of the waveform. The only way you can get it on the

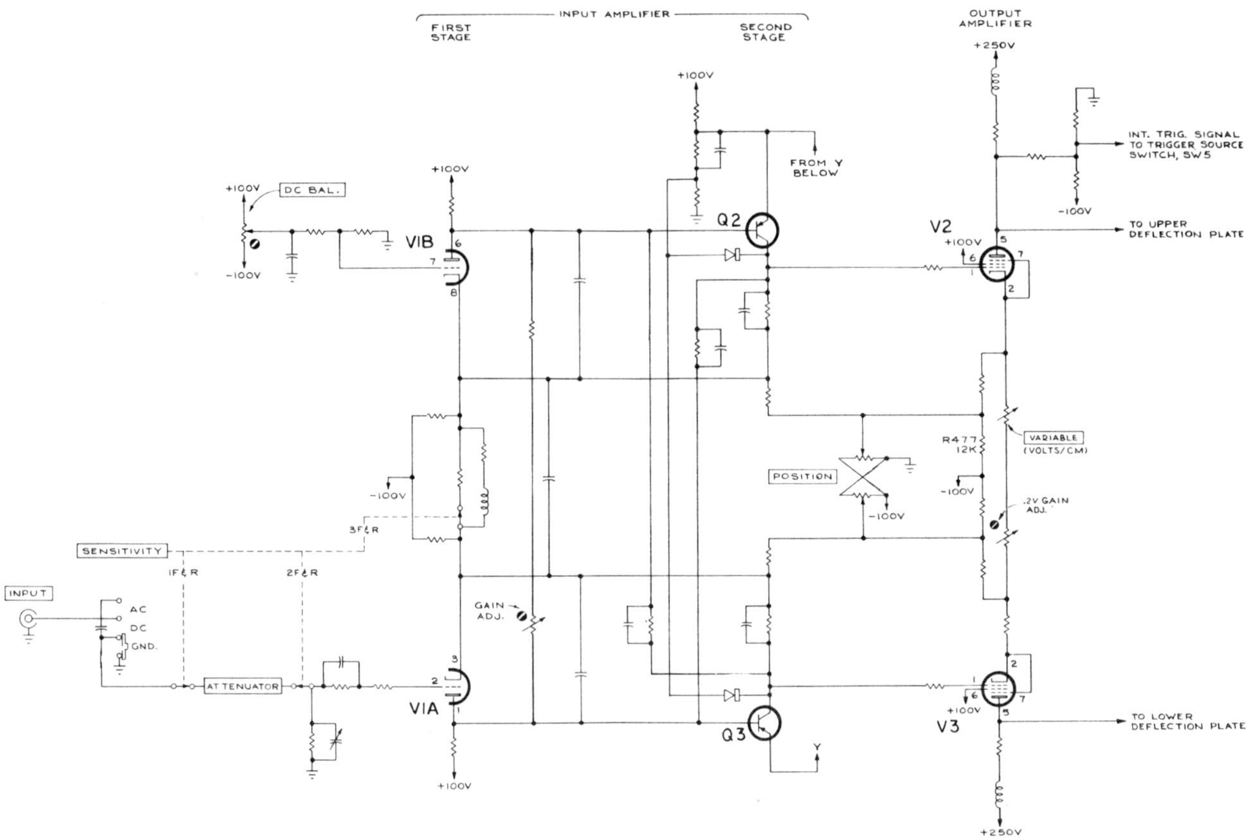

Figure 3.6 Vertical amplifier with single-ended input and push-pull output. (Tektronix RM 504.)

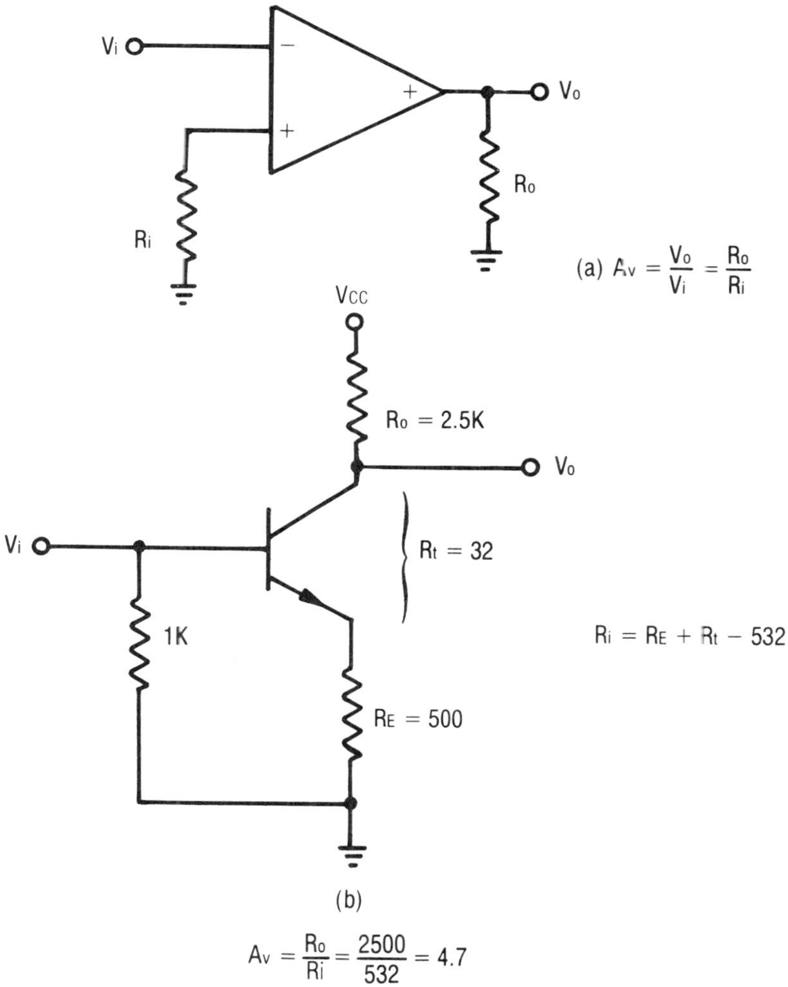

Figure 3.7. Figuring amplifier voltage gain.

screen is by using a slower sweep and displaying two waveforms.

If this is the best our scope can do we'll be limited to quite low-frequency measurements. We won't be able to make accurate measurements of the risetimes of fast pulses at all.

This is shown in Figure 3.8, where we want to measure the pulse risetime of a one-megahertz pulse train. If our scope has a delay line and a one-nanosecond-per-division time base, we can get a display like that in (a), which enables us to determine very accu-

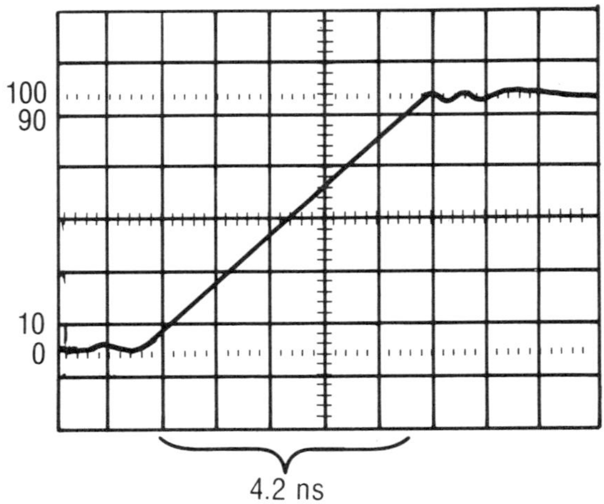

(a) Scope with triggered sweep set at 1 ns/div lets you measure time interval between point on leading edge of pulse where amplitude is 10% and point where it is 90% because delay line shows arrival of pulse leading edge at vertical deflection plates until after sweep gets started.

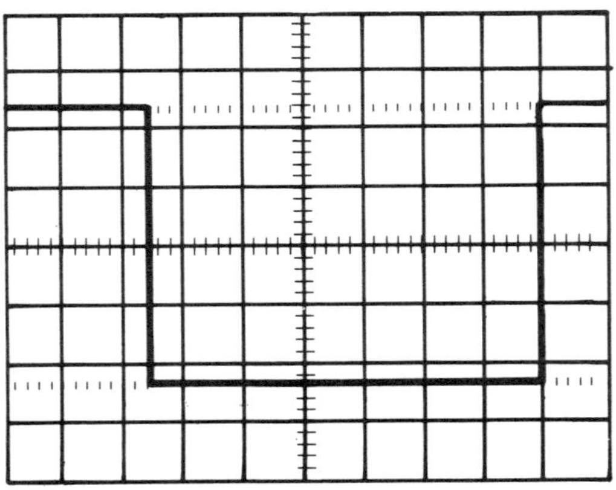

(b) In a scope with no delay line you cannot see leading edge of first pulse, and leading edge of second pulse is too steep at 0.1μ s/div to measure rise time.

Figure 3.8. Risetime measurement.

rately that the time measured along the horizontal scale between the 10-percent level and the 90-percent level of the leading edge is 4.2 nanoseconds.

But if we do not have a delay line we shall not be able to see the leading edge of the pulse because the first 80 nanoseconds of the pulse have gone before the sweep starts. The largest display of the leading edge we can get is that in (b) where we can see the leading edge of the second pulse at the right of the screen. To get this display, we had to use a much slower sweep of 0.1 microsecond per division. This is 100 times slower than the sweep in (a), and you can see it is quite impossible to measure the risetime. Furthermore, this portion of the display falls too close to the margin of the screen, which is not recommended for precise measurement.

We'll come back to risetime measurement and its importance later. We brought it up here to illustrate the how and why of delay lines.

A delay line is a parallel-wire or coaxial transmission line. Its *characteristic impedance* Z_c is given by:

$$Z_c = L/C$$

We find in most scopes the best value for Z_c is about 1000 ohms. This impedance must be constant all along the line. If there is a variation in Z_c at any point, some of the signal's energy will be *reflected,* or transmitted in the reverse direction, from this point, which is called a *discontinuity*. A discontinuity could be caused by a change in the spacing between the conductors, for example. We want to avoid this because it will distort the signal.

If we terminate the delay line with a load impedance equal to Z_c, and drive it with a high-impedance amplifier output circuit, we get a signal voltage on the deflection plates that is twice the output voltage of the amplifier. This is the same as if the vertical amplifier's amplification had been increased by 6 dB. The same will happen if the output impedance of the amplifier is equal to Z_c and the load impedance is made very high.

Figure 3.9(a) shows the arrangement of the components of a *lumped parameter delay line*. It is composed of cascaded LC networks. This line will show the risetime of all signals down to about 10 or 12 nanoseconds. For shorter risetimes we have to have a different kind, a special type of coaxial cable (Figure 3.9(b)).

This line is a specially braided cable in which two insulated conductors are wound spirally in opposite directions on an insulator

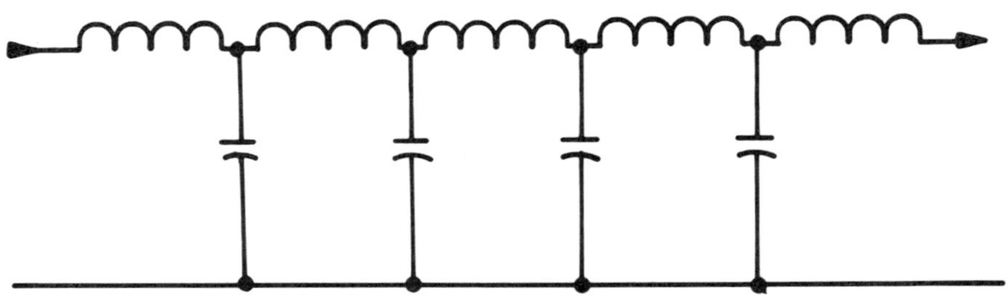

(a) Lumped-constant line is made up of discrete components.

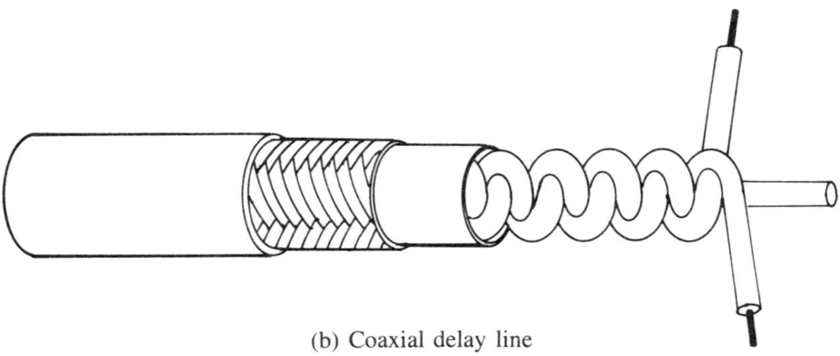

(b) Coaxial delay line

Figure 3.9. Delay lines

core. They are then enclosed within a dielectric and a braided metal outer shield as in common coax. But whereas it would take some 150 feet of common coax to give a delay of about 200 nanoseconds it takes only about 4 feet or so of special coax to do the same.

In the coaxial delay line the inductance and capacitance are inherent in the line instead of being "lumped" in discrete components. A coaxial delay line therefore is called a *distributed delay line*.

This type of delay line may be found in the middle of the vertical amplifier instead of after the output. Wherever the delay line is located it is obvious that the trigger pickoff must precede it.

We mentioned earlier that many scopes are built to use *plug-ins*. These scopes are called ''plug-in'', ''laboratory'' or ''modular'' instruments. The kind of scope that does not use plug-ins is called a ''monolithic'' or ''dedicated'' scope.

The plug-in scope is much more versatile than the built-in monolithic, which is dedicated to one particular function. Plug-ins enable you to change the scope's function by changing its amplifiers. In the simpler models, only the vertical amplifier or preamplifier is a plug-in unit, the other circuits being permanently built-in. The more elaborate units, however, may have as many as four plug-ins—two for the vertical circuits and two for the horizontal circuits—leaving only the CRT and power supplies to be provided in the *mainframe*.

The most popular vertical amplifier in both plug-in and monolithic scopes is the *dual-trace amplifier* or preamplifier, which has two identical amplifier channels. There is a separate input connector for each channel, so that two different input signals can be applied at the same time. Both channels are connected to the CRT, and an electronic switch turns them on and off alternately. Each signal is displayed in turn on the CRT, but a fast switching rate makes it seem as if both are being displayed simultaneously and continuously.

This method should not be confused with the *dual-beam oscilloscope* which is really two oscilloscopes in one, sharing a CRT screen, but otherwise having separate vertical amplifiers and deflection systems. The dual-beam scope can display two unrelated signals, using different sweep rates. The dual-trace amplifier operates with one horizontal deflection system, so obviously the signals displayed must be frequency-related. However, this system is less expensive and perfectly satisfactory for a great number of applications.

A typical dual-trace amplifier has two input connectors, two AC/DC switches and two VOLTS/DIV controls. The two channels may be used separately or together, as selected by a MODE switch. This switch usually has settings designated A ONLY, B ONLY, ALTERNATE, CHOPPED, ADDED ALGEBRAICALLY (see Figure 3.10).

The A ONLY and B ONLY positions select that channel and cut the other off. The switch controls a multivibrator, holding it in

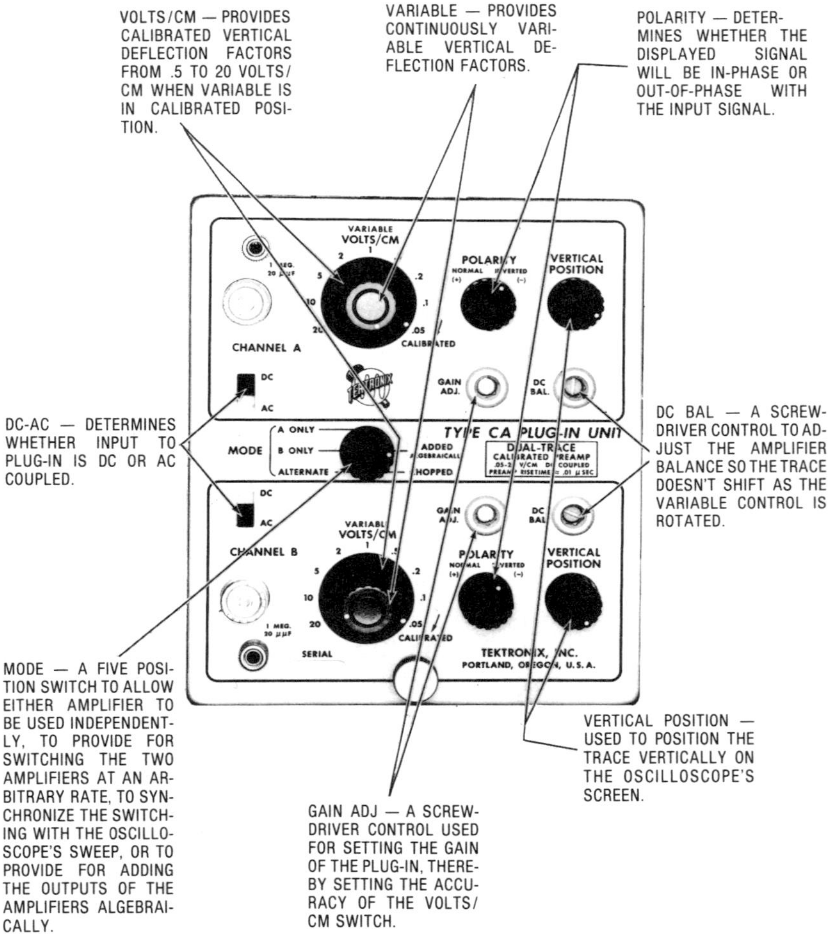

Figure 3.10 (a). Dual-trace pre-amplified plug-in. (*Reproduced by courtesy of Tektronix, Inc. All rights reserved.*)

one of its states, during which its two outputs turn one channel on and the other off.

In the ALTERNATE position a trigger pulse from the horizontal deflection circuit causes the multivibrator to switch to its other state at the end of each horizontal sweep. This turns A channel on and B channel off for one sweep, then A channel off and B channel on for the next, and so on. The switching rate will be according to the sweep rate, so this mode is not suitable for very slow sweeps because the traces alternate instead of both being on the screen at the same time.

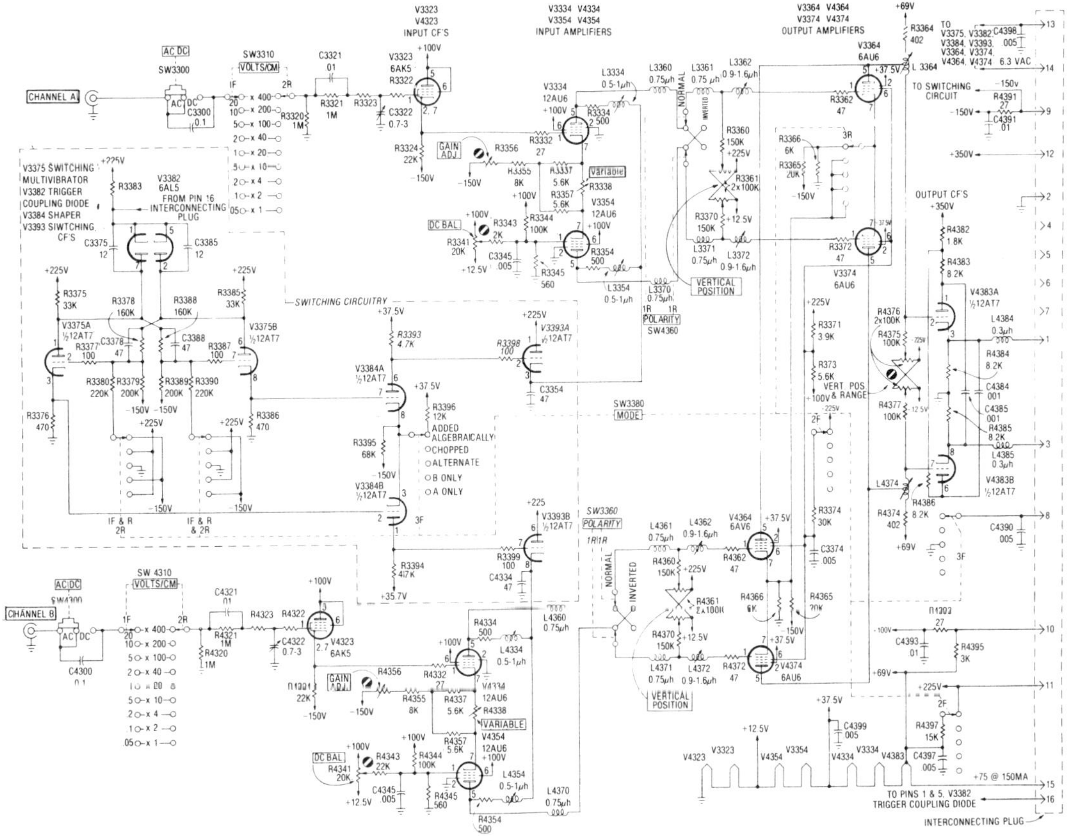

Figure 3.10 (b). Dual-trace preamplifier schematic *(Courtesy Tektronix, Inc.).*

In the CHOPPED position the multivibrator inputs are disconnected from the trigger source and the multivibrator is allowed to free-run at its own frequency. A typical chopped switching frequency is 100 kilohertz.

Rotating the MODE switch to the ADDED ALGEBRAICALLY position places a negative voltage on both grids or bases of the multivibrator, so it is made completely inoperative. This allows both channels to operate at the same time. Signals applied to both inputs will be amplified equally by either channel. If the same signal is present at both inputs the outputs will be equal. Each channel has a POLARITY switch which reverses connections in the channel. If the switch is in its "normal" position the output signal is in phase with the input signal, but if it is in its "inverted" position the output will be of opposite phase. If both POLARITY switches are set the same way, the two outputs will add together to give a signal of twice the amplitude, but if the switches are not set the same way the outputs will subtract. Since they are the same they will cancel each other and the output will be zero.

This last setting enables the plug-in to perform like a *differential amplifier*. This can be a great help in cases when the signal you want to display is superimposed on an undesired signal, such as line frequency hum.

If one input is connected to a source containing both the desired and undesired signals, while the other is connected to a source containing only the undesired signal, you can reduce the undesired signal as much as 20 times by use of the VOLTS/DIV control of the channel containing only the undesired signal.

The VOLTS/DIV control has a "variable" control similar to that on the time-base control. This allows the channel's gain to be varied so that the display can be adjusted to some desired height, but of course it is no longer calibrated. In canceling out an unwanted signal this control acts like a fine adjustment to obtain maximum cancellation.

Whether we have a dual-trace or single-trace amplifier, each amplifier must have a positioning control to move the trace up or down on the screen so you can place it where you want it. It works by changing the d.c. level of the two sides of an amplifier in opposite directions.

To understand this better, look at Figure 3.11, which is a simplified diagram of a single-channel vertical amplifier. In this

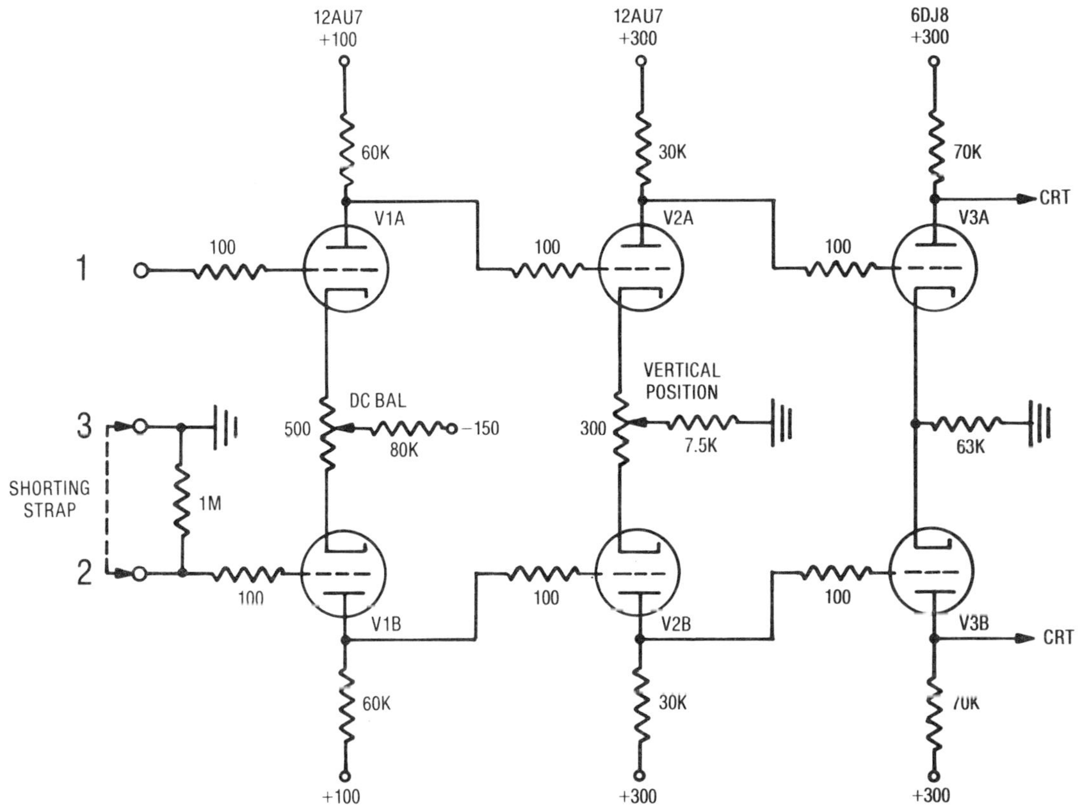

Figure 3.11. Simplified single-channel vertical differential amplifier.

circuit the vertical position control works by changing the cathode resistances of V2A and V2B in opposite directions. This naturally changes the grid-cathode bias, and so increases or decreases the plate current in each tube. The result is that the plate voltage rises or falls accordingly, and this produces an increase or decrease in the d.c. level on the grids of V3A and V3B.

Most vertical amplifiers are differential amplifiers. As we mentioned above, this enables them to eliminate unwanted signals by canceling those that are common to both amplifiers. For that reason such signals are called *common-mode signals* and the amplifier's ability to reject them is called *common-mode rejection*.

Another reason for using differential amplifiers is because the stages are d.c. coupled. This is essential, of course, where you have a positioning control or any other control that operates by changing the d.c. level. D.c. coupling also enables the amplifier to handle signals with very low frequencies. However, d.c. amplifiers are prone to d.c. drift, which is undesirable, but, since it is usually common to both channels, will be canceled out in a differential amplifier.

The differential amplifier in Figure 3.11 has been designed so that it can be converted to a push-pull amplifier by grounding one of the inputs. This changes the first stage (V1A, V1B) into a paraphase amplifier similar to that in Figure 5.8 in Chapter 5.

Requirements of High-Frequency Amplifiers

Low-frequency amplifiers are comparatively simple, and are generally similar to that shown in Figure 3.6. Where the bandwidth is less than one megahertz a delay line is unnecessary, and the trigger signal can be taken from one of the deflection plates, which it often is.

In high-frequency amplifiers the story is quite different. The capacitive reactance to high-frequency signals of the distributed capacitance in the circuit is very low. In the low-frequency amplifier, distributed capacitance of 50 picofarads will offer 12.7 kilohms of reactance to a 250-kilohertz signal, but the same distributed capacitance offers only 12.7 ohms of reactance to a 250-megahertz signal. In other words, it practically grounds it.

Since some distributed capacitance cannot be avoided, how can we build vertical amplifiers to handle these ultra-high frequencies? There are various ways, some of which you already know about because they are used in TV receivers.

The use of peaking coils, for instance, is a means of turning the distributed capacitance of a circuit into an asset. In a circuit such as that in Figure 3.12, L1 and L2 are peaking coils. L1 forms a parallel-resonant circuit with the distributed capacitance so as to offer increased resistance to higher frequencies. Without L1, higher-frequency signals would increasingly bypass the load resistor, so that they would decrease the gain of the amplifier for them. L1 is, therefore, chosen to have an inductance value that has the opposite effect. Because it is in parallel with the distributed capacitance of the circuit, what it does is called *shunt peaking*.

L2, on the other hand, forms a series-resonant circuit (*series peaking*) with the input capacitance of the following stage. Its ability to transfer signal energy therefore *increases* as it approaches resonance. Between them, L1 and L2 decrease signal loss at higher frequencies. To cover as wide a range as possible, they are tuned to different frequencies and provided with resistors to broaden their bandpass by lowering their Q.

In scopes for higher frequencies every effort is made to reduce the input capacitance to the vertical amplifier. Whereas in low-

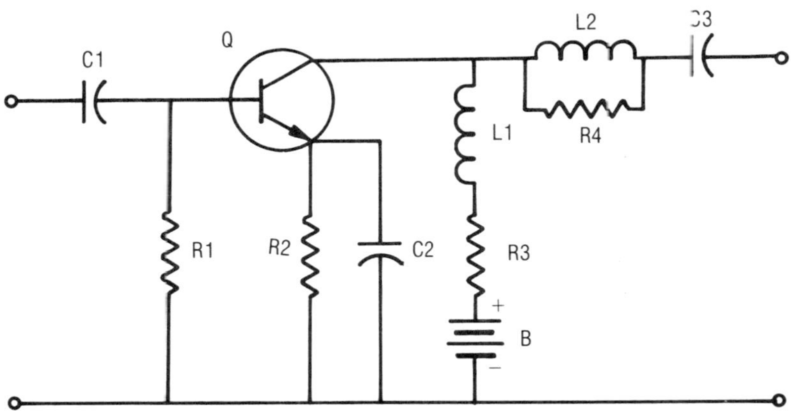

Figure 3.12. Peaking coils are used to extend bandwidth

frequency instruments the value is not critical, in those handling high frequencies it must be kept as low as possible. Where the desired bandwith is 250-megahertz, for example, the input capacitance would have to be 5 or 6 picofarads. To get such low input impedances, careful design of the input is necessary, including even the connector (which is generally the small BNC type). You will never see banana-plug jacks on a high-frequency scope! Resistors and capacitors must also be chosen to provide the proper RC combination that will maintain the high-frequency response of the input.

The input attenuator is a frequency-compensated voltage divider. At d.c. and low frequencies the resistors in the network do all the work, but at higher frequencies the reactance of the capacitors decreases, so that it becomes primarily a capacitance voltage divider. Each attenuator circuit contains an adjustable series capacitor that is set to provide the best response for the high-frequency components of the signal, and an adjustable shunt capacitor for optimum response for the lower-frequency components. These must all maintain the same input impedance regardless of how they are combined for different settings of the volts/division switch.

Peaking by either RC or LC networks is provided through most stages of a vertical amplifier. In particular, peaking networks will be found in the stage following the delay line, because however well you design a delay line, it cannot respond equally to all frequencies, due to skin-effect losses that become worse as the frequency increases.

Another feature found in some high-frequency scopes is the use of a CRT with a *distributed deflection plate system*. In this type of CRT, instead of just a pair of vertical deflection plates there are two sets (upper and lower), connected as in Figure 3.13. The signal passes through the inductors inside the CRT at the same rate as the electrons in the beam passing between them. This synchronization between the electrons in the circuit and those in the beam allows the signal electrons to influence their opposite numbers in the beam for a longer period (the transit time between the system of plates) and so results in an increased deflection. This effect is of great help at the higher frequencies, where signal fluctuations are too swift for the conventional deflection system, resulting in serious reduction of high-frequency sensitivity.

Calibrator

Before making amplitude measurements of a displayed signal you should always make sure that the vertical amplifier gain is adjusted accurately. You do this by applying a known, peak-to-peak, voltage to the vertical input and observing the display. For instance, if we use a signal with a voltage of one volt peak-to-peak, and the vertical VOLTS/DIV control is set to 1, we'd expect to see the signal displayed with an amplitude of exactly one screen division. If we then found that another displayed signal had an amplitude of, say, 1.5 divisions we could state with confidence that its amplitude was 1.5 volts peak-to-peak.

Most scopes have a calibrating-signal source built in. This varies from a front panel jack that provides a single voltage to a more elaborate circuit that enables you to select any one of as many as 18 different voltages, from 0.2 mV to 100 V peak-to-peak. The voltages are generally in the form of a square-wave signal with a

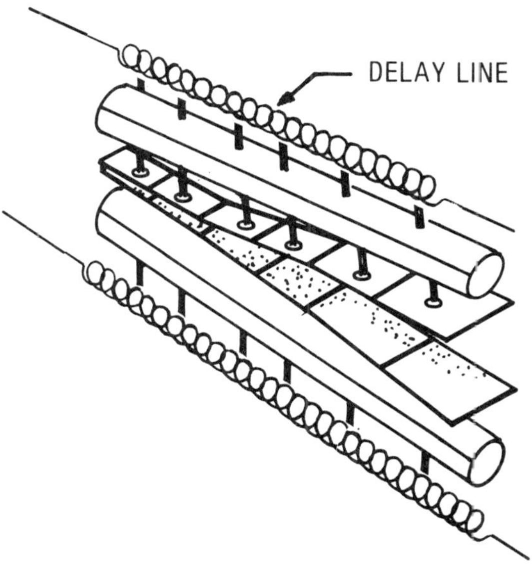

Figure 3.13. Distributed deflection-plate system

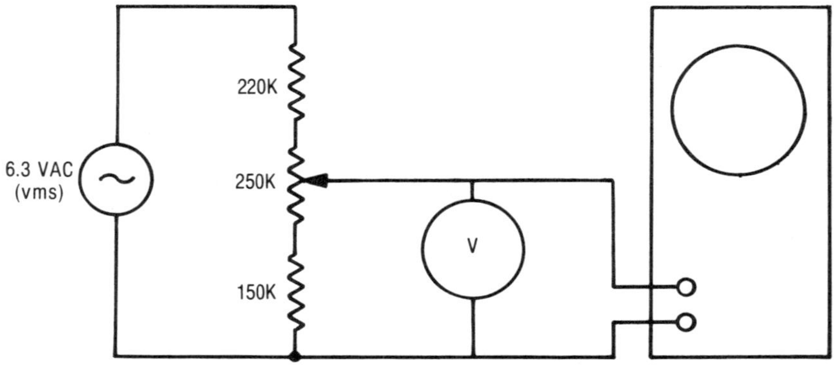

Figure 3.14 Calibration device.

frequency of one kilohertz. In some cases two frequencies are available, one kilohertz and one megahertz.

The accuracy of the calibrator amplitude is extremely important, of course, and in good-class modern scopes is usually within one percent of the nominal value.

Those scopes without a built-in calibrator can, of course, be calibrated by applying a signal from an external source to the vertical input. A square wave at about one kilohertz is best, but a 60-hertz sine wave from the scope's power supply will do at a pinch. This may need to be reduced by means of a resistive voltage divider, and must also be monitored by a reasonably accurate voltmeter. Bear in mind that most voltmeters read the root-mean-square value of a sine wave—but *not* of other waveforms—and that your scope will display the peak-to-peak value, so you must convert the rms value to peak-to-peak (multiply it by 2.828).

It is seldom necessary to use more than one calibrating signal to calibrate the vertical amplifier. If you make for yourself a little calibration device with a circuit similar to that in Figure 3.14, you can adjust the potentiometer to give a full-scale deflection on the scope of, say, 1 volt per division. On a screen with six vertical divisions this would be six volts peak-to-peak, or eight volts for eight divisions. These would read 2.12 or 2.83 V rms respectively on the rms voltmeter if the signal is a sine wave.

If the voltmeter reads differently, you should reset the potentiometer for the proper value, and then adjust the vertical amplifier gain adjustment to obtain the correct deflection.

After that, switch to some of the other ranges to check that the deflection is still correct for those settings of the volts/division control.

4

Generating the Sweep

We most often use our scopes for displaying the way in which some quantity varies with time. The variations of the quantity are applied in the form of analogous voltages to the vertical deflection plates in the CRT. As we saw in the previous chapter, we can measure the resultant beam deflections on the vertical graticule scale, or Y axis.

The time interval between different points in the display is measured along the horizontal, or X axis. While the vertical plates deflect the beam up and down, the horizontal deflection plates must make it sweep from left to right across the screen, at a steady rate and at a convenient speed, to give a useful display.

At the beginning of each sweep the voltage on the left-hand plate goes high and that on the right-hand plate goes low. This attracts the electron beam toward the left-hand plate, so that it strikes the left-hand side of the screen. The left-hand plate potential then decreases while the right-hand plate potential increases. The strength of attraction exerted by the left-hand plate declines as its voltage goes down, while that on the right-hand plate rises. The electron beam moves from left to right accordingly.

It is highly desirable that the electron beam moves across the screen at a constant speed. The time it takes to pass each division of the graticule must be the same, or time measurements (and consequently frequency measurements) will be distorted. The sloping portion of the waveforms of Figure 4.1 should be straight, so the change of voltage on the horizontal deflection plates will be linear. In modern scopes this has been achieved over at least 80 percent of the sweep. Some minor distortion in the first and last of the ten

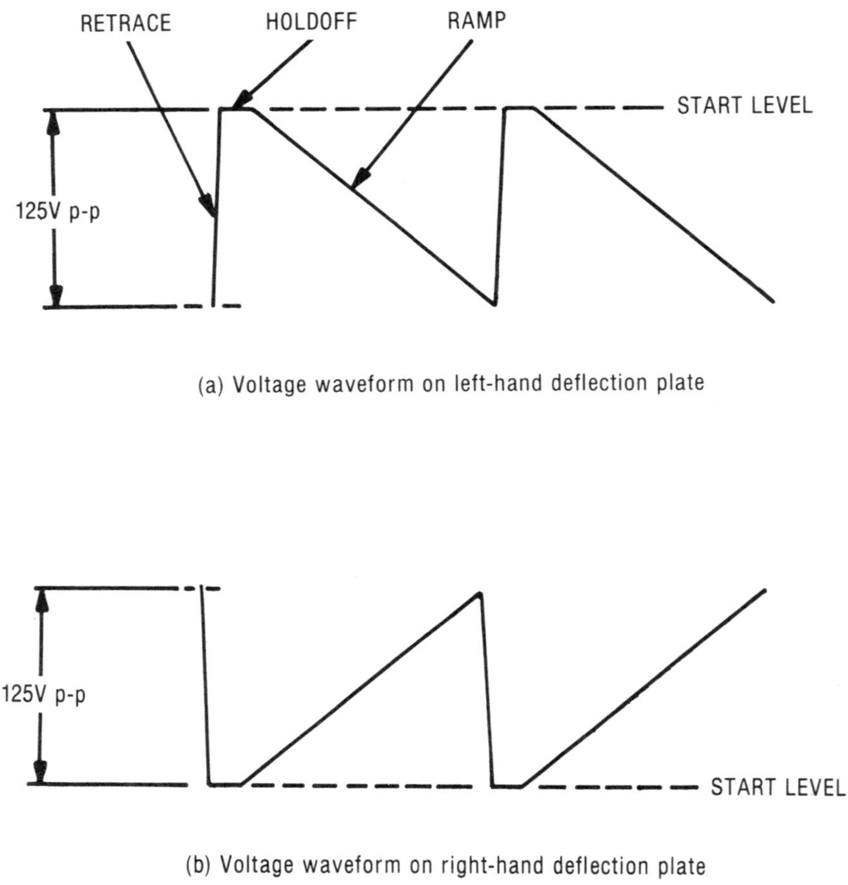

Figure 4.1 Typical horizontal deflection-plate potentials. Voltage variations for two successive triggered sweeps are shown. The actual potentials will not be the same in different scopes and CRT's but their "profiles" will be alike.

horizontal divisions may exist, mainly due to CRT characteristics. The chain of circuits from input to CRT is shown in Figure 4.2. The heart of the horizontal deflection system is the *time-base generator,* which we'll discuss first.

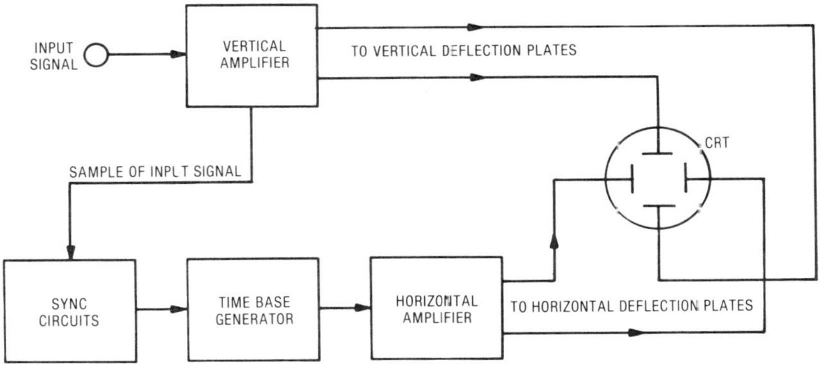

Figure 4.2 Circuits driving the CRT deflection plates.

The time-base generator is a sawtooth oscillator that has to produce a sloping ramp voltage similar to Figure 4.1(a) or (b). Its single-ended output is coupled to the horizontal amplifier, which has a push-pull output driving the CRT horizontal deflection plates.

All time-base generators are based upon the circuit shown in Figure 4.3. When we close the switch S in this circuit the capacitor

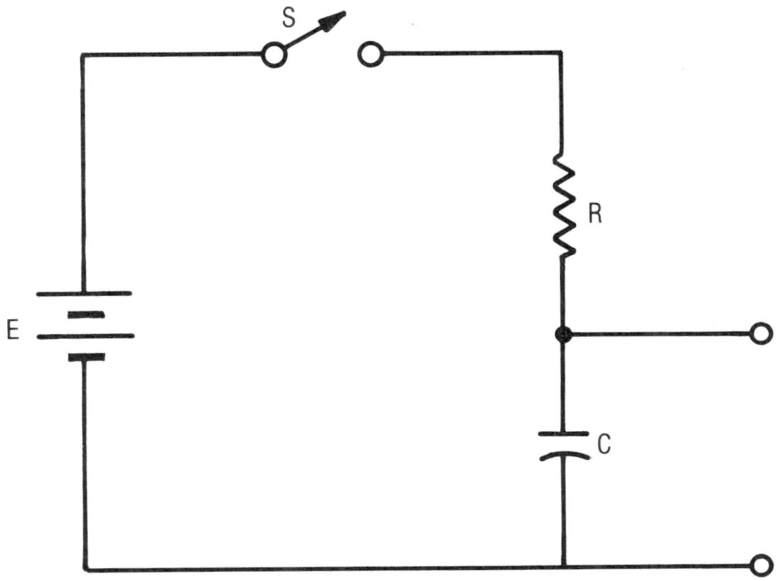

Figure 4.3. Basic Time-Base Generator Circuit

C charges up to the battery potential E. The rate at which it charges gradually decreases because the increasing capacitor voltage opposes the battery voltage. This progressively reduces the charging current until it finally stops altogether, as shown in Figure 4.4.

If it were not for the increasing opposition of the capacitor voltage, the charging rate would follow the dashed line OE′, and the capacitor would be fully charged after an interval equal to its capacitance multiplied by the value of the resistance R (Figure 4.3). For example, if R is 1 megohm and C is 0.01 microfarad,

$$T = (1 \times 10^{6}) \times (0.01 \times 10^{-6}) = 0.01 \text{ s, or } 10 \text{ ms}$$

This is the *RC constant* of R and C in series, and is independent of the battery voltage. However, because of the slowing down of the charging rate the capacitor voltage in practice reaches only 63.2 percent of E when T = RC.

All sweep generating circuits are designed to get better linearity than that in Figure 4.4. The simplest circuits do this by using only the first ten percent of the curve, as in Figure 4.5. This can be done with a circuit such as that in Figure 4.6. This circuit uses a thyratron, which is in a non-conducting state (turned off) as current from the power supply flows through R to charge C. When C's potential reaches V's firing point, V switches to its conducting state (turned on) and discharges C. V turns off again and the cycle is repeated. The firing point of V is set by the bias on V's grid.

The frequency of operation of this "relaxation oscillator" depends on the RC constant of R and C. If R is made variable, the frequency can be varied over a range, and if there are several C's selectable by a switch there can be several ranges, as in Figure 4.7.

In operating a scope with this type of sweep you select a suitable range for the frequency of the signal you want to observe, and then adjust the variable control until a stable display is obtained. This is synchronized by feeding a portion of the signal to the input of the thyratron circuit. When the variable control is set so the circuit's free-running oscillation frequency is just slightly less than that of the signal, the two lock together. This is because the thyratron is almost at its firing point when the next positive-going excursion of the signal arrives on its grid and, by adding to the bias voltage, triggers it into conduction. When it fires, the plate voltage of the thyratron goes almost instantly to zero. The circuit is designed

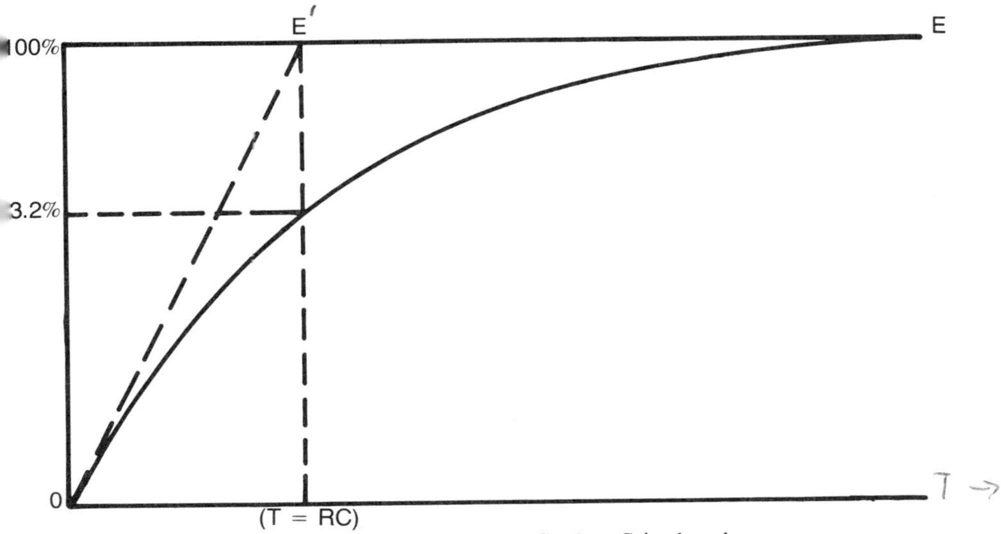

Figure 4.4 Graph of voltage on C when S is closed.

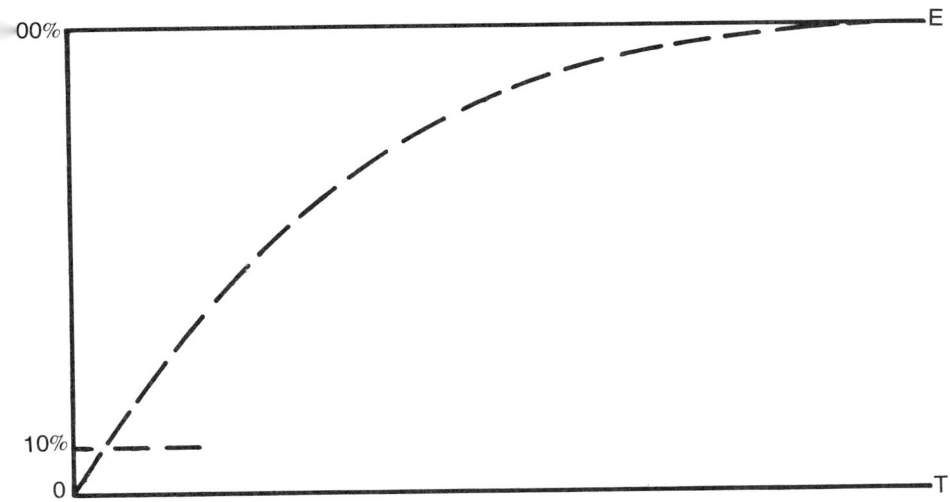

Figure 4.5 Portion of curve of Figure 4.4 used in exponential sweep to obtain better linearity.

so that this results in conduction being cut off again, so the RC combination begins another charging cycle, as in Figure 4.8(a).

If the free-running oscillation frequency of the thyratron circuit is set to slightly under half that of the synchronizing signal it will be

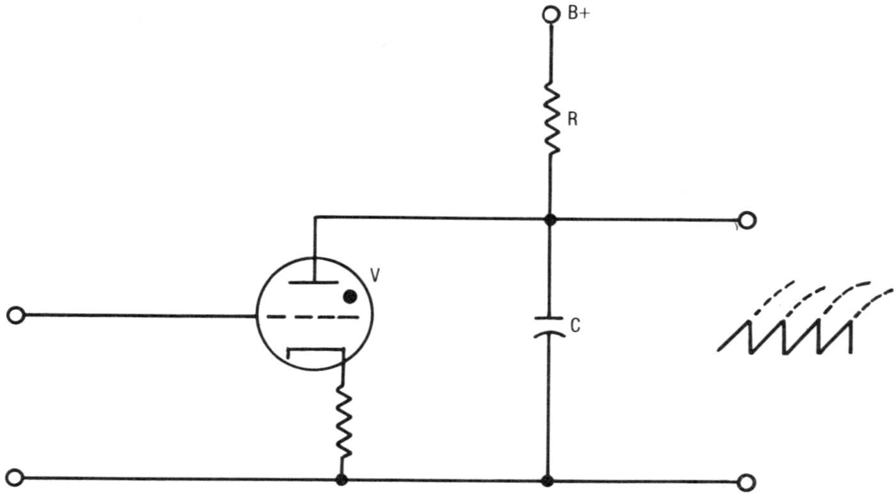

Figure 4.6 Thyratron sweep circuit

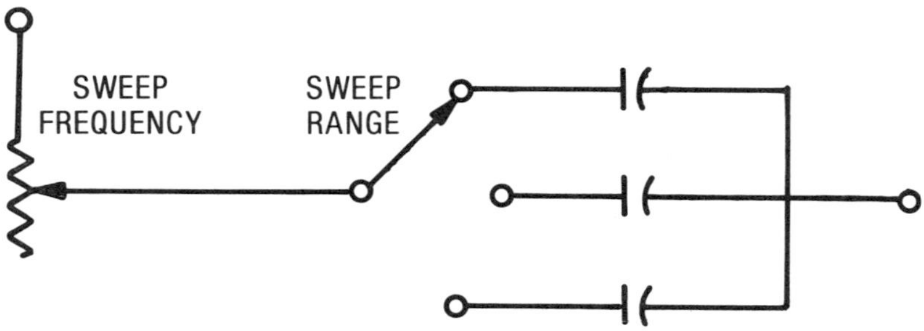

Figure 4.7 Sweep frequency control

ready to be triggered into conduction only on every other cycle of the signal. This results in a display of two cycles, as shown in Figure 4.8(b). Similarly, if you adjust the frequency to one-third you get three cycles, and so on.

However, if you set the free-running frequency to twice that of the synchronizing signal's frequency, the display will consist of one half of the signal on the first sweep and the other half on the second, as in Figure 4.8(c). Faster sweeps produce more and more complicated displays. Consequently, this type of sweep synchronization is

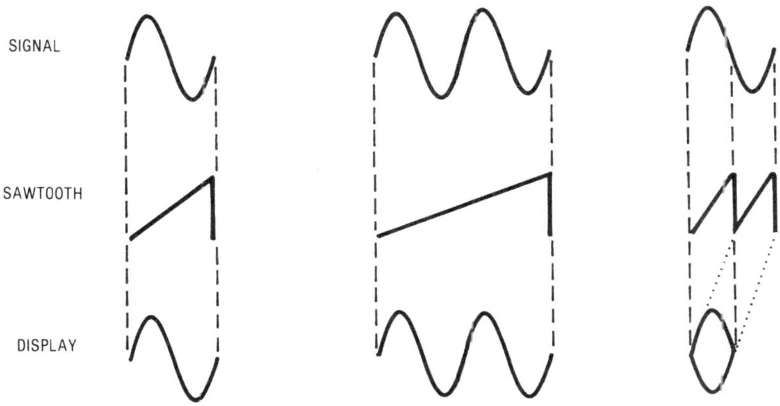

(a) Sweep frequency same as signal frequency. (b) Sweep frequency half signal frequency. (c) Sweep frequency twice signal frequency.

Figure 4.8 How sweep is synchronized.

not suitable for sweep speeds that are faster than the frequency of the signal being displayed.

This circuit is not used in modern scopes. Similar circuits, such as blocking oscillators and multivibrators employed in TV receivers, work in the same manner, and give the same results as long as the ramp consists only of the first 10 percent of the RC exponential curve. In a scope these circuits must be variable over a full range of sweep frequencies, which is not the case in TV receivers, of course.

The amplitude of the synchronizing signal is adjusted by means of the *sync amplitude* control. This control is necessary because the amplitude of the input signal may be high and if it is not reduced will result in an unsynchronized display. Starting with this control set too low, you adjust it to obtain a display that locks in satisfactorily, but do not advance it beyond that point or the display loses sync again.

In the absence of an input signal the sweep still runs at its own frequency, and the CRT screen shows a straight horizontal line. This is now a free-running sweep, not controlled by a synchronizing signal. Because it always runs continuously, whether synchronized or not, and because each sweep begins almost instantly after the preceding sweep stops, this type of sweep is called a *recurrent sweep*.

The recurrent sweep has been superseded in most modern scopes by the *triggered sweep*. In this type the sweep runs for an exact length of time, during which it is independent of the frequency of the input signal. A portion of the signal is applied to a trigger circuit which converts it to pulses of uniform amplitude and shape. Equal pulses prevent variations in sweep-circuit operation caused by changing input signals. On receipt of a pulse the sweep starts and runs for the set time you select. If other trigger pulses arrive while it is running, they are ignored. After the sweep has been completed it starts again with the next pulse, which it waits for. In this way it is synchronized with the input signal, yet independent of it as well. With this sweep you can display small portions of the waveform by increasing the sweep speed without having it fold back over itself as in Figure 4.8(c).

The sweep generator is turned on and off by a gating signal which comes from a previous circuit called the *sweep-gating multivibrator,* located between the trigger circuits and the sweep generator (see Figure 4.9). There are several types of sweep generator circuits controlled in this way. We shall describe three of them:

Gated-clamp-tube generator
Bootstrap sweep generator
Miller integrator sweep generator

The *gated-clamp-tube generator* and sweep-gating multivibrator circuits are shown in Figure 4.10. The trigger pulse produced by the preceding trigger circuits is a negative-going pulse with a waveform resembling that at A. It arrives on the cathode of V1, making it negative with respect to the plate. V1 is therefore biased on, and transmits the pulse via C2 to the grid of V3.

V2 and V3 are a monostable multivibrator. In its resting condition V2 does not conduct, but V3 does. The arrival of the negative pulse on V3's grid turns this tube off, so V3's plate voltage rises immediately toward the value of E_{bb}. This causes a positive-going voltage to appear on V2's grid, turning this tube on. Its plate voltage promptly falls practically to zero.

This negative-going gating voltage (waveform B) is coupled to the control grid of V4, causing this tube to switch off. It remains in this condition as long as the gating voltage remains on its grid. Meanwhile C5a or C5b, the timing capacitor selected by the range switch S1, charges through R7b, a variable timing resistor, resulting in the ramp voltage (waveform C). This appears on V4's plate and is coupled via C3 to the input of the horizontal amplifier.

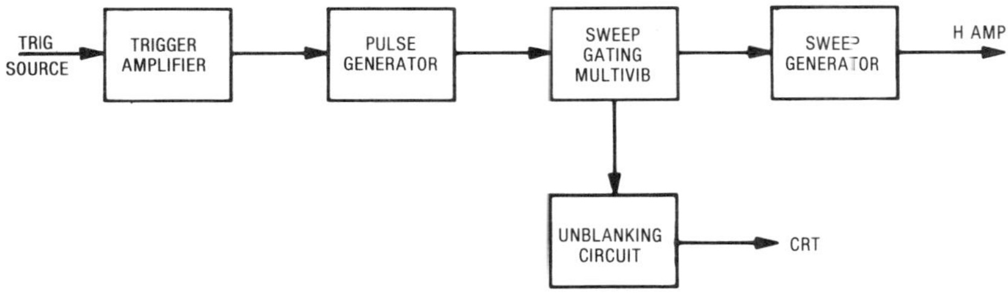

Figure 4.9 Triggered sweep block diagram

V5 is normally conducting, since its plate is at zero volts while its cathode is slightly negative. This results in the output being clamped at the level of its cathode at all times except when the positive-going ramp reverse-biases it and cuts it off. When the ramp drops back to its low level, V5 turns on again and ensures that the next ramp starts at the same level.

Variable resistor R7b is ganged with R7a. R7a is part of an RC combination consisting of R7a, R6, C2 and V2 (which has *some* resistance while conducting).

As soon as V2 starts conducting C2 begins charging from E_{bb}. The charge on C2 appears also on the grid of V3 (which is cut off). As soon as the potential on V3's grid becomes high enough to turn V3 on, its plate voltage falls sharply. This negative-going impulse is transmitted via C4 to V2's grid, and V2 is turned off. The sweep-gating multivibrator is now back to its original state. The gating voltage disappears from the grid of V4, and this tube conducts and discharges C5a or C5b, as applicable, thereby terminating the ramp. R7a and R7b are matched together to ensure that the gating voltage pulse and the sawtooth ramp each has the same time interval.

R5 is a *stability control*. This control sets the bias on V2's grid. Although the sweep-gating multivibrator is operated in the monostable mode to obtain triggered sweeps, it is possible to obtain a free-running sweep by adjusting R5. You often need to see where the trace is when there isn't a signal to trigger it, for example, so you can position it on a certain horizontal graticule line to establish a baseline for amplitude measurements.

This circuit has some drawbacks. To begin with, to get a 10-percent sawtooth of practical size we need a power supply ten times as great. A sawtooth with a peak voltage of 150 volts would therefore require a power supply of 1500 volts. On the other hand, if we

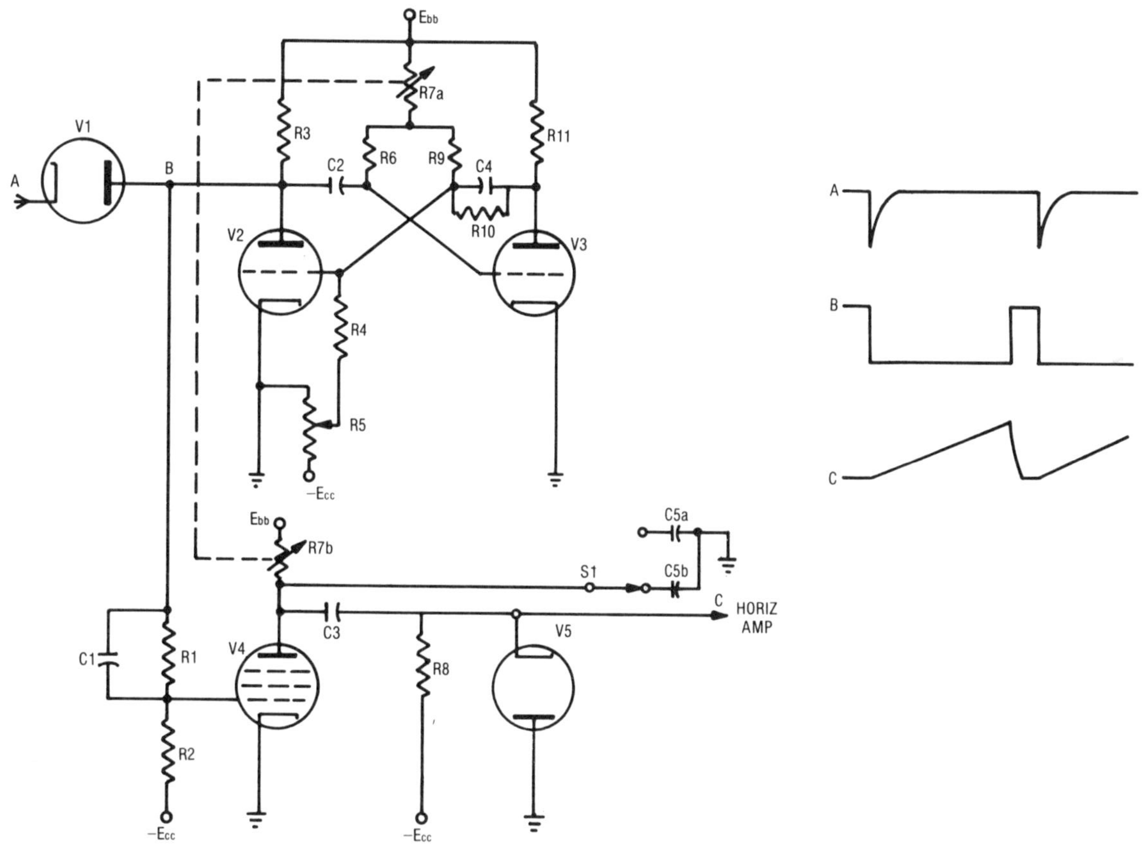

Figure 4.10 Gated-clamp-tube sweep generator and sweep-gating multivibrator

limit E_{bb} to, say, 250 volts, we can get a sawtooth of only 25 volts, requiring a horizontal amplifier to drive the CRT horizontal deflection plates. This also limits the sweep speeds we can get.

Another limitation on sweep speeds comes from the RC coupling between V4 and V5, which imposes a low-frequency cutoff. This prevents the circuit from generating slow sweeps.

The sweep linearity is still not as good as we'd like for really accurate timing, and there is no way to control or adjust the sweep length. Furthermore, it is difficult to maintain the fixed ratio between the timing components in the sweep generator and the sweep-gating multivibrator that is essential. Finally, there is no *hold-off* at the end of each sweep to give the circuits a chance to make a full recovery before the next sweep.

The problem with obtaining a sawtooth from an RC circuit, as we have seen, is that the increasing opposition of the charge on the timing capacitor opposes the charging current. Even the first 10 percent of the exponential current curve is not linear enough for really accurate time bases. However, if we could find a way of maintaining the charging current at a fixed rate we'd get a perfect ramp.

One way to do this would be by adding a circuit that increases the supply voltage in proportion to the increasing charge on the timing capacitor so as to maintain a constant charging current.

Figure 4.11 shows a circuit designed to do this. Before the gating voltage is applied to its grid, V1 is biased at zero volts. Maximum current flows from the low side of the circuit through V1, R2 and V2 to E_{bb}. The potential on the plate of V1, and therefore on C1, is virtually zero also as the tube is in its fully-conducting state.

When the gating pulse cuts V1 off, the timing capacitor C1 begins to charge toward E_{bb}. Since it is connected to V3's grid this rises by the same amount. But the more V3's grid moves in a positive direction, the more heavily V3 conducts. This results in an increasing current through R3, which in turn gives a rising potential at the cathode of V3 and also on C2. If V3 has a loop gain of 1, the rise of potential on C1 and C2 will keep in step. However, a point will be reached when their combined voltage will exceed E_{bb}, so V2 will be reverse biased and will not conduct any more. Now C2 becomes the power supply for C1. The charging current now travels from the low side of the circuit via V1, R2, C2 (as this is not d.c. it

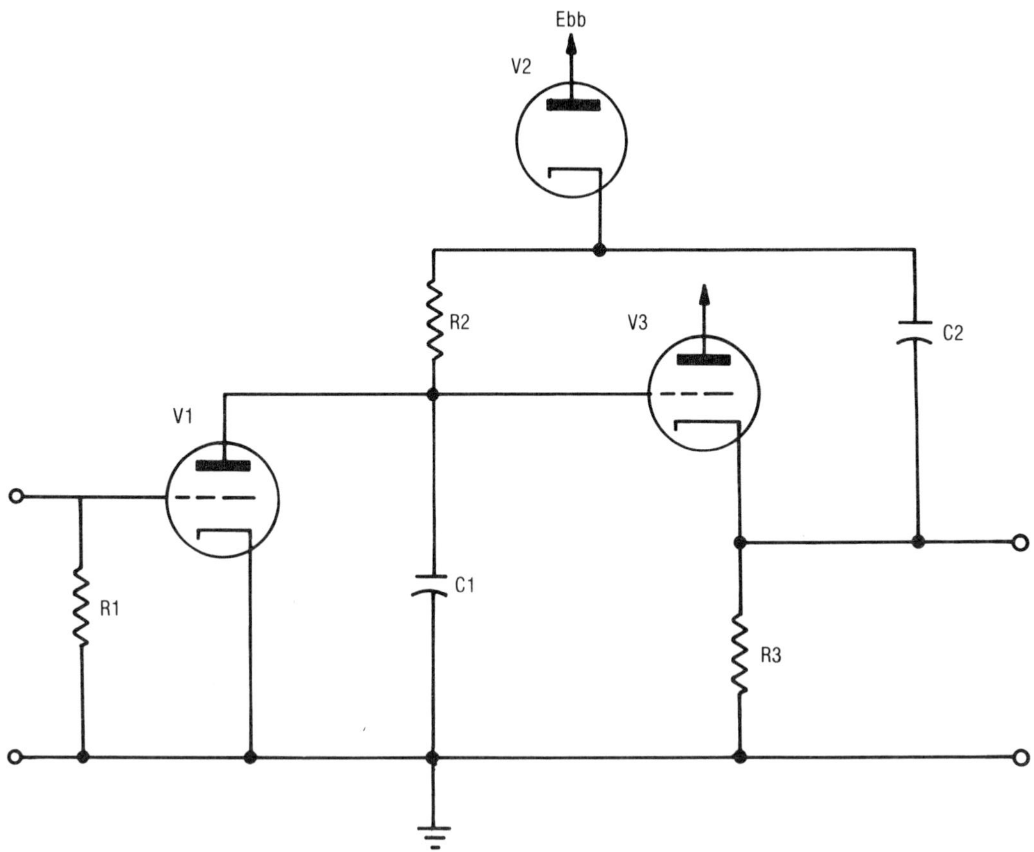

Figure 4.11 Bootstrap sweep generator

can pass through C2), and V3 to its plate supply. As V3's plate supply is considerably higher than E_{bb} this charging current continues while V1 is cut off. When the negative gate voltage is removed from V1's grid and the tube resumes conducting, C1 is discharged almost instantly, and V2 becomes forward biased again and discharges C2. The circuit is now ready for another cycle of sweep generation.

The positive feedback just described increases the voltage on V2's cathode at a rate that maintains a constant voltage drop across R2 so the charging current is constant. But this will happen only if the total gain around the feedback loop is unity. If it is not, the voltage on V2's cathode will not maintain a constant voltage drop across R2 and a non-linear ramp will be generated.

The circuit consisting of V3, R3 and C2 is called a *bootstrap circuit,* so this sweep generator is called a bootstrap generator. It overcomes many of the disadvantages of the clamp-tube generator. It generates a sawtooth with an amplitude large enough to drive the CRT horizontal deflection plates directly, which not only is an economy, but also allows the use of much faster sweep speeds.

However, tube aging and line voltage variations will affect the linearity. Use of a pentode rather than a triode for V1 is preferable because a pentode has a more constant current. Line voltage variations affect tube heater temperatures even if the plate voltages are regulated, so a special regulated heater supply is desirable for the sweep generator. The capacitance of C2 must also be about 20 times greater than C1 to prevent non-linearity due to the tendency of the charging current to discharge C2.

The principle of the *Miller integrator time-base generator* is illustrated in Figure 4.12. At first C1 has no charge. Then we close the switch S1. Now electrons flow from B− through R1 to the lower side of C1. Electrons also flow from the upper side of C1 through BT1 and R2 to B+. As the charge on C1 increases it opposes the charging current, which as we've seen already is the usual case. Consequently, the charging current begins to fall off. However, this falling off results in a decrease of the voltage drop across R1, and V1's grid becomes more negative. Because of this the current through V1 and R2 diminishes, and the voltage drop across R2 lessens. As a consequence, V1's plate voltage rises, so that the voltage on the upper side of C1 becomes more positive. This brings

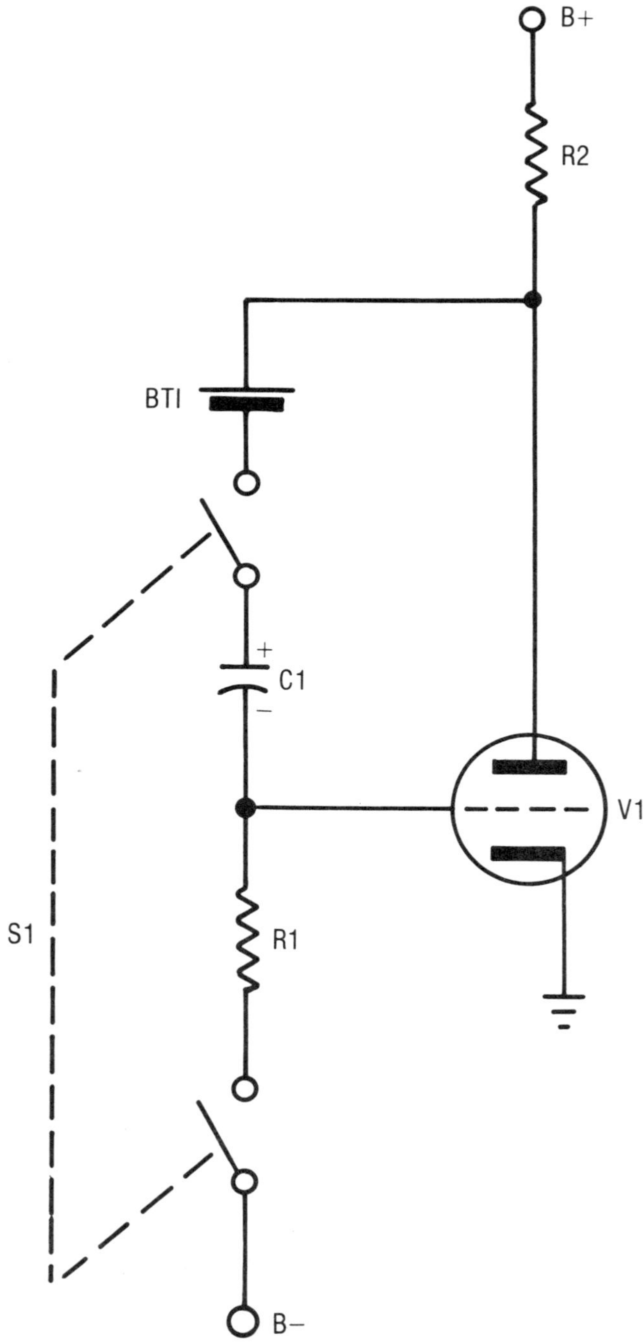

Figure 4.12 Principle of Miller integrator sweep generator.

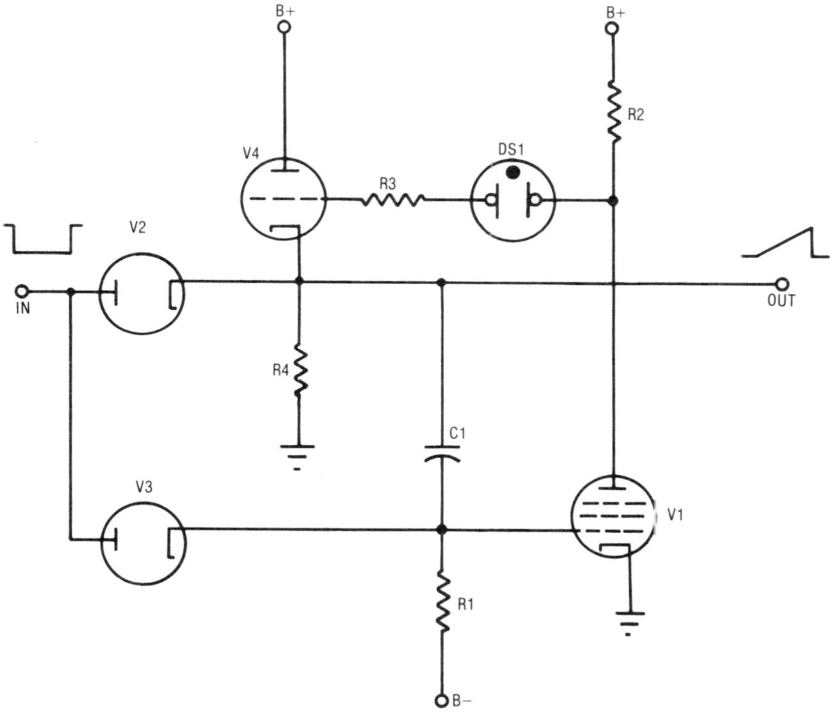

Figure 4.13 Basic Miller run-up circuit

the charging current back to its original value. In this way the charging current is kept linear.

The battery BT1 is included in the circuit to prevent V1's plate and grid voltages being the same. If they were, V1 would not operate normally. In a real circuit we'd use a different method.

For instance, in the circuit in Figure 4.13 the battery has been replaced by the neon lamp DS1. The constant voltage across neon lamps maintains a difference of 50 volts between their electrodes. This in turn keeps the grid of V1 50 volts lower than the plate, regardless of how the voltage of the latter rises. The electron path this time is from the upper side of C1 via V4 to B+. As the plate voltage on V1 rises the grid voltage of V4 rises. DS1 ensures it is always 50 volts lower. This increasing grid voltage causes the electron flow from ground via R4 and V4 to B+ to increase, so that the voltage drop across R4 rises, producing a linear ramp at the output.

This ramp voltage also appears on the upper plate of C1 with the effect described above.

V4 is a cathode follower, so it also has the effect of isolating the runup circuit from any shunting effect from the following circuit that would degrade its performance. The cathode circuit of V4 is clamped by V2 to provide a fixed reference voltage for the ramp to start from and return to. It shares the gating signal with V3, which is responsible for turning the ramp on and off.

In the absence of the gating signal both V2 and V3 are conducting, because the input resting voltage makes their plates slightly positive compared to their cathodes. While V2 is conducting, the cathode of V4 and the output of the sweep generator share the voltage existing at the cathode of V2 resulting from the electron flow through R4 and the diode. The bias on V4's grid at this time is such that most of the current in R4 is due to V2, so the output voltage returns to this level after each sweep. This ensures that the sweep will always start at the same point on the screen, with no sideways jitter.

V3 also conducts, so that current from B− flowing through R1 continues via V3 instead of charging the lower plate of C1. This keeps V1's grid at a constant level, similar to that of the cathode of V4.

When the gating signal arrives at the input it makes the plates of V2 and V3 more negative than their cathodes. The two diodes now cannot conduct. They are like two switches that have been turned off. For this reason they are called *disconnect diodes*.

The current that previously flowed through V3 now starts to charge the lower plate of C1. This makes the grid of V1 increasingly more negative. Conduction through V1 diminishes so its plate voltage rises. This causes V4's grid to move in a positive direction, therefore the current in this tube increases also. The voltage drop across R4 rises, so the voltage on V4's cathode rises too.

This rising voltage appears at the output and also on the upper plate of C1 as electrons flow from it via V4 to B+. As already explained, this linearizes the output of V1, and consequently the output of V4.

This circuit is called a runup circuit because the ramp rises in a positive direction. You can also have a Miller rundown circuit, with the ramp going in a negative direction. It doesn't make any differ-

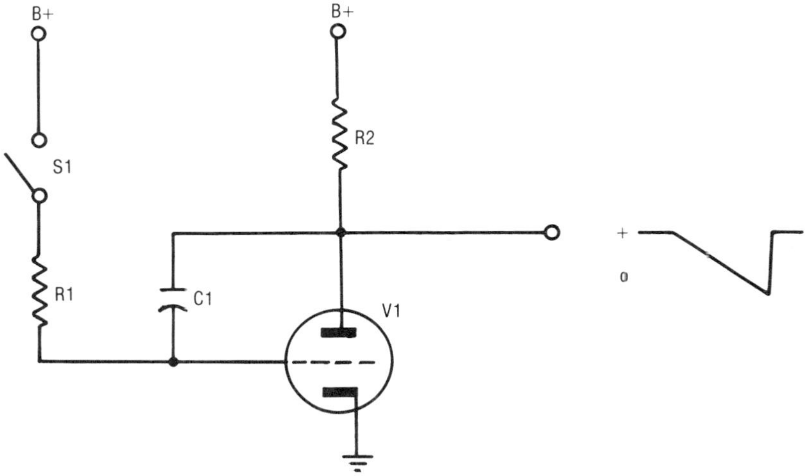

Figure 4.14 Principle of Miller run-down sweep generator
With S1 open the upper side of timing capacitor C1 is at the same potential as the plate of V1, so that a considerable potential difference exists between the upper and lower plates. When S1 is closed the lower plate also starts to be positively charged via the timing resistor R1. As the positive voltage increases, the current through V1 and R2 increases, causing a larger voltage drop across R2. This lowers the potential on the upper side of C1 and at the output as shown by the waveform. Should the current in R1 fall off the lessened voltage drop causes the voltage on V1's grid to fall. This makes the current through V1 and R2 increase so V1's plate voltage falls faster to offset the lessening current in R1. In this way the linearity of the ramp is preserved.

ence, because the output of the sweep generator is coupled to a paraphase amplifier in the horizontal amplifier, where it is split into two signals of opposite phase to drive a push-pull output.

Figure 4.14 shows the basic principle of a Miller rundown circuit.

A Miller rundown sweep generator that also generates its own gating signal is shown in Figure 4.15. It is called a *phantastron sweep generator*. A trigger signal is applied at either input 1 or input 2. If input 1 is used, the trigger must be negative. A positive trigger is required at input 2.

At first the plate voltage of V1 is high, its level fixed by the clamp tube V2. The voltage on the suppressor grid, however, is low

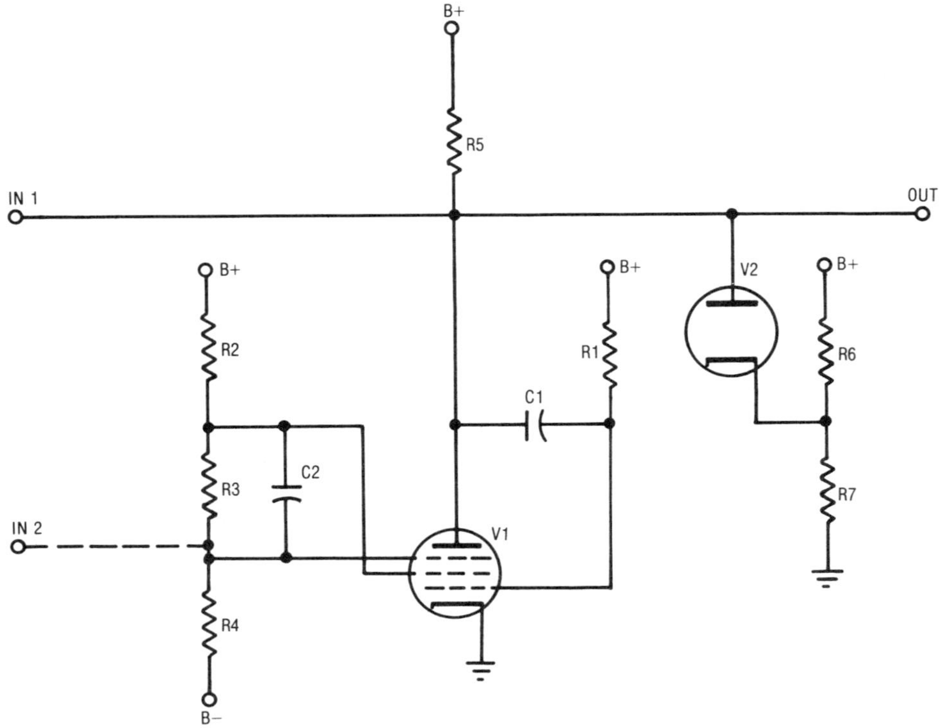

Figure 4.15 Phantastron sweep generator

so that no plate current flows. Nevertheless, the screen grid draws heavy current because the control grid is well above cutoff. The voltage on the screen grid is low because of the voltage drop across R2 caused by the screen current.

To generate a sweep, we apply a negative-going trigger pulse at input 1. This is coupled via C1 to V1's control grid, which is driven below cutoff. The tube accordingly stops conducting and the screen current ceases. This causes the screen grid voltage to rise sharply, and this positive-going pulse is coupled via C2 to the suppressor grid. It now allows plate current to start flowing. The plate voltage drops and lowers the potential on the left-hand plate of C1. The right-hand plate now begins to discharge through R1 to B+. Since C1 and R1 are the timing capacitor and resistor, a rundown ramp is developed as in the Miller rundown circuit shown in Figure 4.14.

Eventually, the voltage on V1's plate gets so low that it does not attract as many electrons as before, and the screen grid takes over again. The circuit then immediately returns to its original state, ready for the next trigger pulse.

As mentioned above, a positive trigger can be applied at input 2. This is coupled directly to the suppressor grid, starting plate current flow. The resultant plate voltage drop is coupled via C1 to V1's control grid, where it reduces the screen current, causing a rise in screen voltage. The rest of the action is the same as for the negative-going trigger already described.

In a practical sweep generator, R4 is a variable resistor, called the *stability control*. When adjusted to a very low resistance value it keeps the suppressor grid voltage so negative at all times that the trigger pulse cannot override it to start the sweep, so it does not run. On the other hand, if it is rotated to where the resistance value is high, the suppressor grid's voltage never gets low enough to turn V1 off, so a free-running sweep results. At some intermediate position the stability control allows the suppressor grid to synchronize with incoming trigger pulses as described above.

Two more requirements that must be taken care of in a gated sweep generator are *hold off* and *unblanking* (the latter is a feature of all modern sweep generators). The hold-off circuit feeds back a portion of the output sawtooth waveform to the input of the sweep-gating multivibrator. When it reaches the proper voltage it returns the multivibrator to its original state, removing the negative gating voltage from the sweep generator input, so that the rundown (or runup) ceases. However, the holdoff circuit delays the feedback signal by passing it through an additional RC circuit so that it cannot trigger the multivibrator until the sweep generator has returned to its resting state. This is done to ensure that each sweep starts from the same point. If it did not, it would not appear directly on top of its predecessor, and the display would be blurred.

The unblanking circuit applies a positive voltage to the CRT control grid while the sweep is running. This allows the electron beam to be emitted from the cathode to trace the display on the screen. A portion of the gating signal is used, so that the CRT is turned on and off in exact synchronization with the sweep. Since the gating signal is often negative, its polarity must be reversed before it

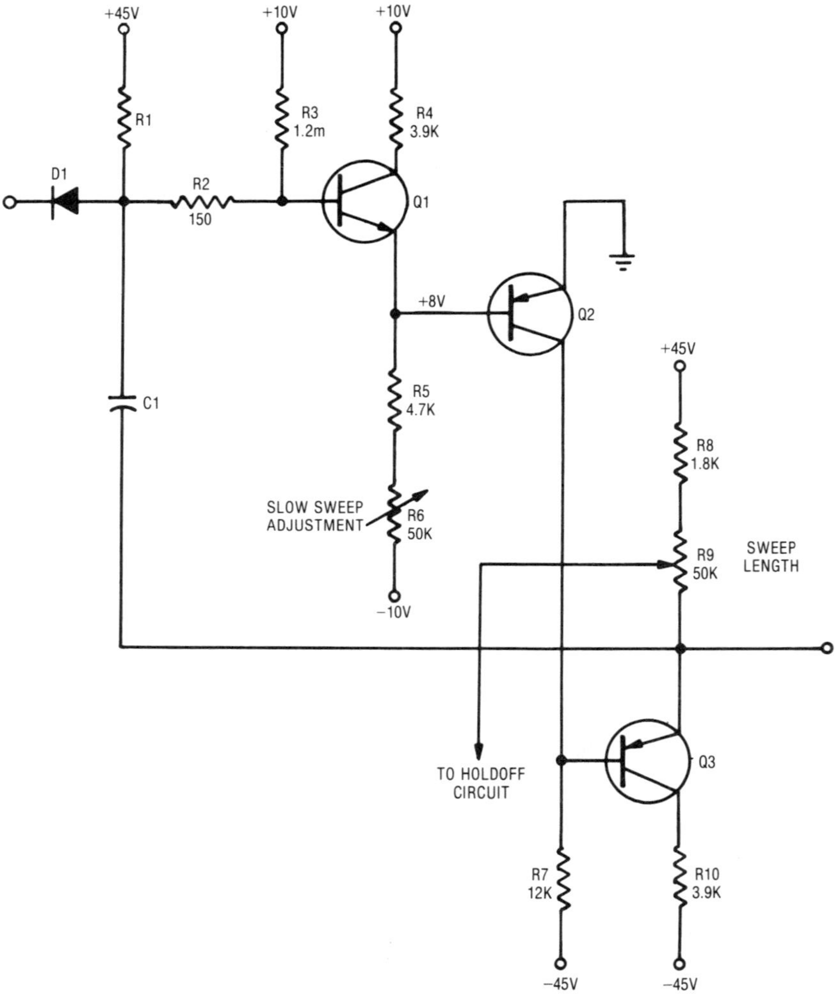

Figure 4.16 Transistor Miller sweep generator

reaches the CRT grid. This is easily done by routing it through a tube or transistor circuit first, which also amplifies it as necessary. On termination of the gating pulse the unblanking signal also disappears and the CRT is cut off until the next gating pulse. This eliminates the appearance of retrace lines on the CRT screen caused by the horizontal deflection plate voltages returning to their initial state at the end of each sweep.

So far we have studied only vacuum-tube circuits used in sweep generators. At first there was some difficulty in designing a transistor version of a Miller integrator sweep generator, due to the low input impedance of transistors. The Miller integrator requires the high input impedance of a vacuum tube to avoid shunting the timing capacitor and resistor, and also the tube's high gain if it is to obtain satisfactory linearity. The circuit shown in Figure 4.16 gets around these problems by adding the emitter-follower transistor Q1 between D1 and Q2 (the disconnect diode and the Miller runup transistor). The combined input impedances of Q1 and Q2 are very considerably higher than that of a single transistor. Emitter-followers also have very high input impedances by themselves, so another, Q3, is used to isolate the sweep generator from the horizontal amplifier to which it is connected.

The base currents of Q1 and Q2 are held constant by R3, which is a high-value resistor. This maintains a constant input impedance that is not affected by variations in the current flowing in R1 and C1.

Earlier in this chapter we mentioned the limitations of the recurrent sweep. The recurrent-sweep generator is a simple sawtooth oscillator that is synchronized directly with the signal being observed, so that one or more cycles are displayed. The rate at which the electron beam scans the CRT screen, therefore, is either the same, or is a definite fraction of the repetition rate or frequency of the signal. Figure 4.7 showed the rather simple frequency controls for this type of sweep generator.

On the other hand, a triggered sweep may be started by a trigger pulse derived from the signal under observation, but it runs at a rate set by fixed resistors and capacitors, and this rate is independent of the signal frequency. The circuit is designed to make the electron beam cross the screen at a fixed rate of so many microseconds, milliseconds or seconds per graticule division. These components are mounted on a rotary switch in such a way that when it is set to a particular sweep rate the proper timing resistor and capacitor are connected into the circuit.

Figure 4.17 illustrates a portion of the time-base timing switch that would be used to connect the proper values of C1 and R1 into the circuit of Figure 4.13. With the switch set as shown, electrons will flow from B− via the 5-megohm resistor to charge the 0.001-microfarad capacitor. The time constant of this combination is:

$$R1 \times C1 = 5 \times 10^{6} \times 0.001 \times 10^{-6} = 0.005 \text{ s}$$

In other words, the electron beam will sweep across the screen in 5 milliseconds. As there are ten graticule divisions this is a rate of 0.5 millisecond per division.

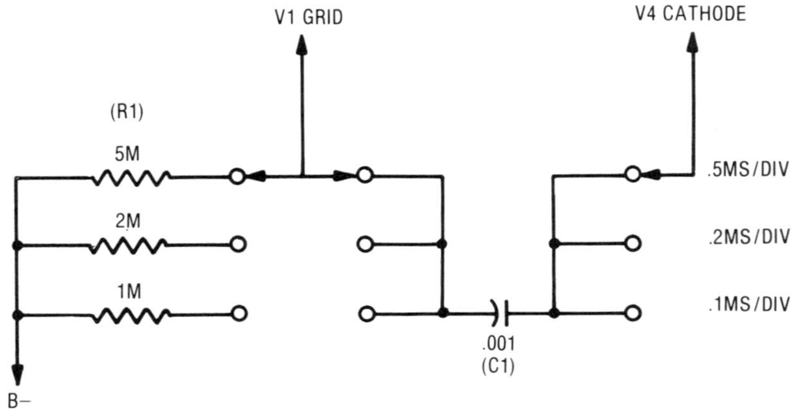

Figure 4.17 Portion of time-base timing switch circuit

By using the same three resistors with capacitors ranging by factors of ten from 1 microfarad down to 100 picofarads we provide sweep rates from 0.5 second to 10 microseconds per division. Higher or lower sweep rates than these are obtained by using smaller or larger resistor values rather than by using extremely large or small capacitors. A typical switch will have settings from 5 seconds to 0.1 microsecond per division in a sequence of 5-2-1 throughout.

The accuracy of these time bases will be no better than that of the components employed. The accuracy is usually within 3 percent of the nominal. This is achieved by using one-percent precision resistors and selected capacitors. Internal calibration adjustments in the form of variable resistors and trimmer capacitors are installed in the time-base circuit to allow for periodic adjustment to maintain accuracy as components age.

Since these time-base rates are fixed, a variable resistor operated by a front-panel control is always provided so you can interpolate between them. The control works in conjunction with a switch that disconnects it completely from the circuit when you are not using it. As long as it is in its disconnected position (marked "calibrated") the sweep rates indicated on the timing switch are true. One of the things you should always check before using a triggered-sweep scope is that this control is in its calibrated position. The knob is usually concentric with the main timing knob, and of a contrasting color.

On the diagram in Figure 4.17 this variable resistor would be switched into the circuit in series between B− and R1.

5

Triggering and Amplifying the Sweep

In Figure 4.8 we saw how a recurrent sweep is started using a portion of the input signal. Although the sync amplitude control allows us some control over the size of the sync signal we have to accept its shape as it is. But not all waveshapes are suitable for this, and in any event we can't select the voltage level or slope of the waveform at which to start the sweep.

A trigger circuit converts the input signal waveform to a pulse of uniform amplitude and shape. With this pulse we can get reliable synchronization with all types of input signal. We can also choose the exact point on the input signal where we want the sweep to start.

In Figure 5.1 you can see the basic parts of the triggering circuit. There are four:

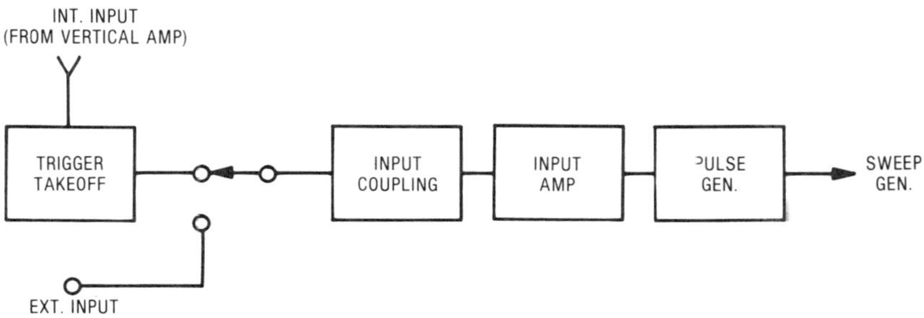

Figure 5.1 Basic parts of trigger circuit

a. *Trigger take-off:*
this has to pass a portion of the vertical amplifier signal to the input-coupling section without changing it unduly, adding noise or other interfering signals. At the same time it must not degrade the performance of the vertical amplifier.

b. *Input coupling:*
this allows you to select different trigger sources. Usually they are named *internal, external* or *line*. The first two may be a-c or d-c coupled. Sometimes low-frequency or high-frequency filters are provided as well.

c. *Input amplifier:*
this amplifier has a high input impedance to avoid loading the vertical amplifier or external input. It is also a comparator in which the input signal level is compared to a set triggering level. A reversing switch allows you to choose the positive or negative slope of the signal.

d. *Pulse generator:*
this is generally a bistable multivibrator that generates pulses in response to inputs from the input amplifier. These pulses are then differentiated to shape them into sharp rising and falling trigger spikes.

The simplest form of trigger take-off would be to use a portion of the signal on one of the vertical deflection plates in the CRT. This avoids the need for additional amplification, and as long as the scope is only to be used for low frequencies the loading of one side of the vertical amplifier doesn't matter. We can improve the frequency response to some extent by using a frequency-compensated voltage divider, and reduce the circuit loading by using a cathode-follower output stage (see Figure 5.2).

High-frequency scopes, however, have a delay line between the vertical amplifier and the CRT. The purpose of this is to give enough time for the trigger and sweep circuits to operate before the signal in the vertical amplifier reaches the CRT deflection plates. Without the delay we couldn't see the first part of a fast signal because it would be gone before the sweep started. In this type of scope we must pick off the trigger signal at some point before the delay line.

As we mentioned above, scopes give you a choice of sources for the trigger signal, of which the most generally provided are "internal," "external" and "line." For most purposes the internal

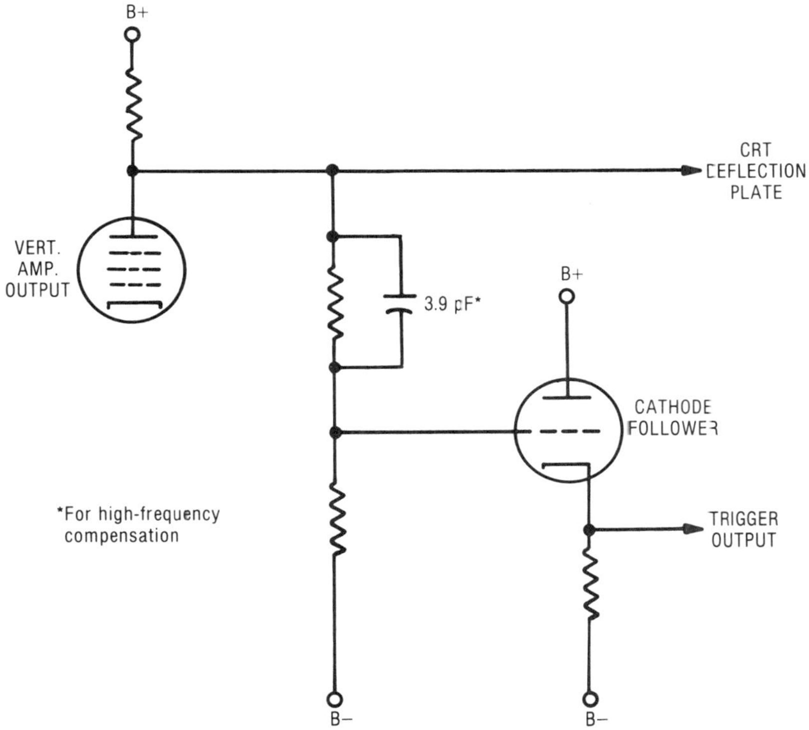

Figure 5.2 Trigger take-off for scope without delay line.

trigger is perfectly satisfactory, especially since the signal has been amplified in the vertical amplifier, whereas the external input jacks provide no amplification. However, external triggering has definite advantages in some cases. For example, if you are viewing waveforms at various points in a circuit you can trigger from the input waveform and compare the time relationships and phases of the waveform at each point to that at the input.

With external triggering the triggering signal also usually remains constant in amplitude and shape, so you don't have to keep resetting the triggering controls for each observation.

You may also want to observe waveforms in power supplies, where a 60- or 120-hertz frequency signal, derived from the power line, exists. Very stable triggering is obtained by using a 6.3-volt tap on the scope's power transformer as the source. Like some external triggers, it has the advantage of constant amplitude and shape.

These sources may be coupled directly or capacitively to the trigger input amplifier. D.-c. coupling is often preferable with low-frequency signals. The triggering-mode control not only enables us to do this, but often has one or two additional positions. A "low-frequency reject" setting provides for a-c coupling through a smaller capacitor which attenuates any low-frequency component, such as hum.

But first we must consider the trigger input amplifier. Figure 5.3 illustrates a typical vacuum-tube type. It is a comparator, in which two tubes share a common cathode resistor. With the switch S1 in the position shown, a set voltage from R7 is applied to V1's grid. The trigger signal goes to V2's grid. R7 is the *triggering level control* and S1 is the *trigger slope switch*.

To understand how this circuit works we must first look at the following circuit which it controls. This is the trigger pulse generator. It is usually a flip-flop, or bistable multivibrator. A commonly used type is the Schmitt trigger (Figure 5.4). In this circuit the voltage divider R1, R2 and R3 sets the grid bias on V2 so that this tube is cut off. V1, however, is conducting, and its plate voltage is low because of the voltage drop across R1.

When a negative-going signal appears on V1's grid, current through the tube decreases and the plate voltage rises. This rise is coupled through C1 to V2's grid, causing the voltage here to go sufficiently positive to allow V2 to conduct. The current through this tube causes its plate voltage to fall. When the negative-going signal is removed from V1's grid the plate current recovers and the plate voltage falls. This has the effect of turning V2 off, so its plate voltage returns to its resting level.

From this description of the Schmitt trigger's operation we can see that it requires a negative-going signal from the trigger-input amplifier.

The trigger-input amplifier in Figure 5.3 is designed so that we can choose whether we want to trigger from the positive or negative slope of the input signal. With the switch S1 set as shown, the input signal goes to the grid of V2. The positive-going half of the signal causes V2 to conduct more heavily so that its plate voltage falls giving a negative-going output.

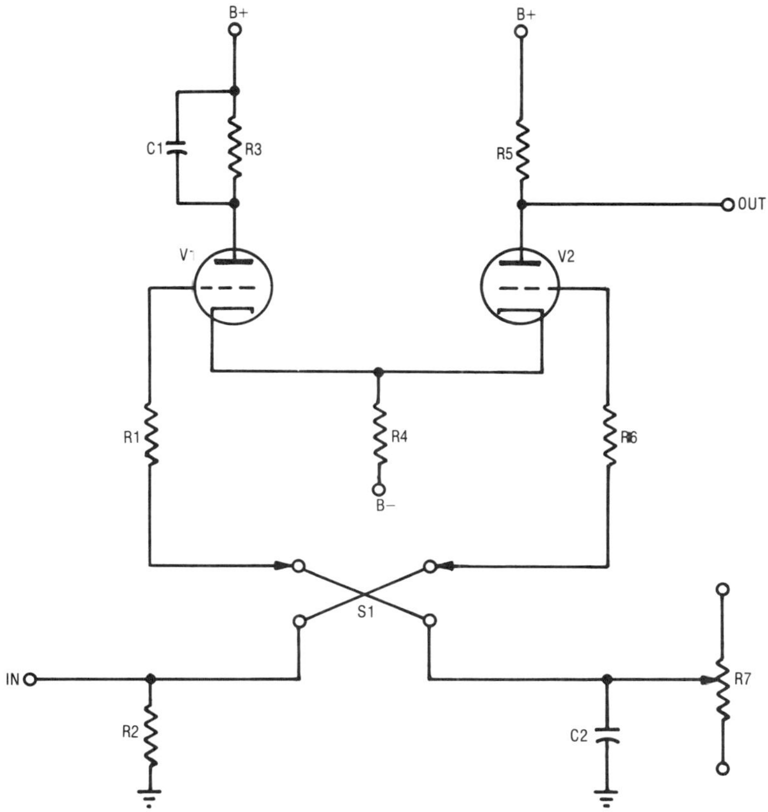

Figure 5.3 Trigger input amplifier

The actual plate voltage of V2 will depend upon the setting of R7. This potentiometer applies a voltage to the grid of V1 that determines the plate current of this tube. This plate current also flows in R4. The voltage dropped across R4 depends upon the strength of the current, so the voltage on the cathodes of V1 and V2 is set by adjusting R7. The cathode-to-grid bias voltage on V2 determines the current in V2 and its plate voltage. Since the Schmitt trigger responds to a certain voltage at its input we can select the exact moment for its operation by the level of the bias on V2. Hence the name "level control" for R7. It selects the point on the positive-going slope of the *input* signal at which the negative-going *output* signal will trigger the pulse generator.

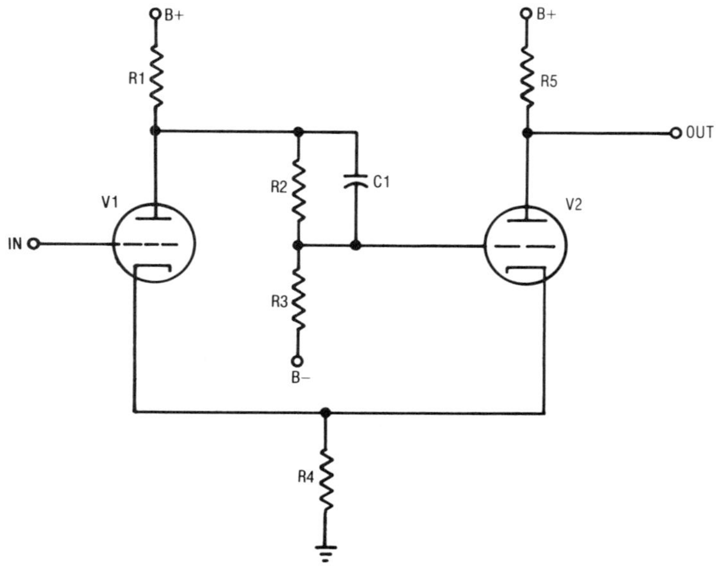

Figure 5.4 Schmitt trigger

If we wish to trigger at a point on the negative slope of the input signal, we set S1 in its other position. The input signal now appears on V1's grid, while the level control is connected to V2's grid. This makes V1 and V2 operate as a flip-flop, so the output at V2's plate has the same phase as the input, and it is still negative-going. The trigger level control now sets the cathode-grid bias of V2 directly, but with the same result in the output signal.

Figure 5.5 shows the same circuit with transistors Q2 and Q3 substituted for V1 and V2. Transistors have a rather low input impedance, however, so the field-effect transistor Q1 is added to avoid loading the vertical amplifier. Diodes D1 and D2 are provided to keep the voltage on Q2's base within safe limits.

Figure 5.6 shows a Schmitt trigger with transistors. Its operation is the same as the circuit in Figure 5.4.

A refinement that may be provided in either of these circuits is a *sensitivity* potentiometer, connected between the two cathodes or emitters, that biases the non-conducting tube or transistor closer to conduction. This increases the sensitivity of the circuit to low-amplitude signals. However, advancing the control too far in the direction of increased sensitivity will make the circuit unstable.

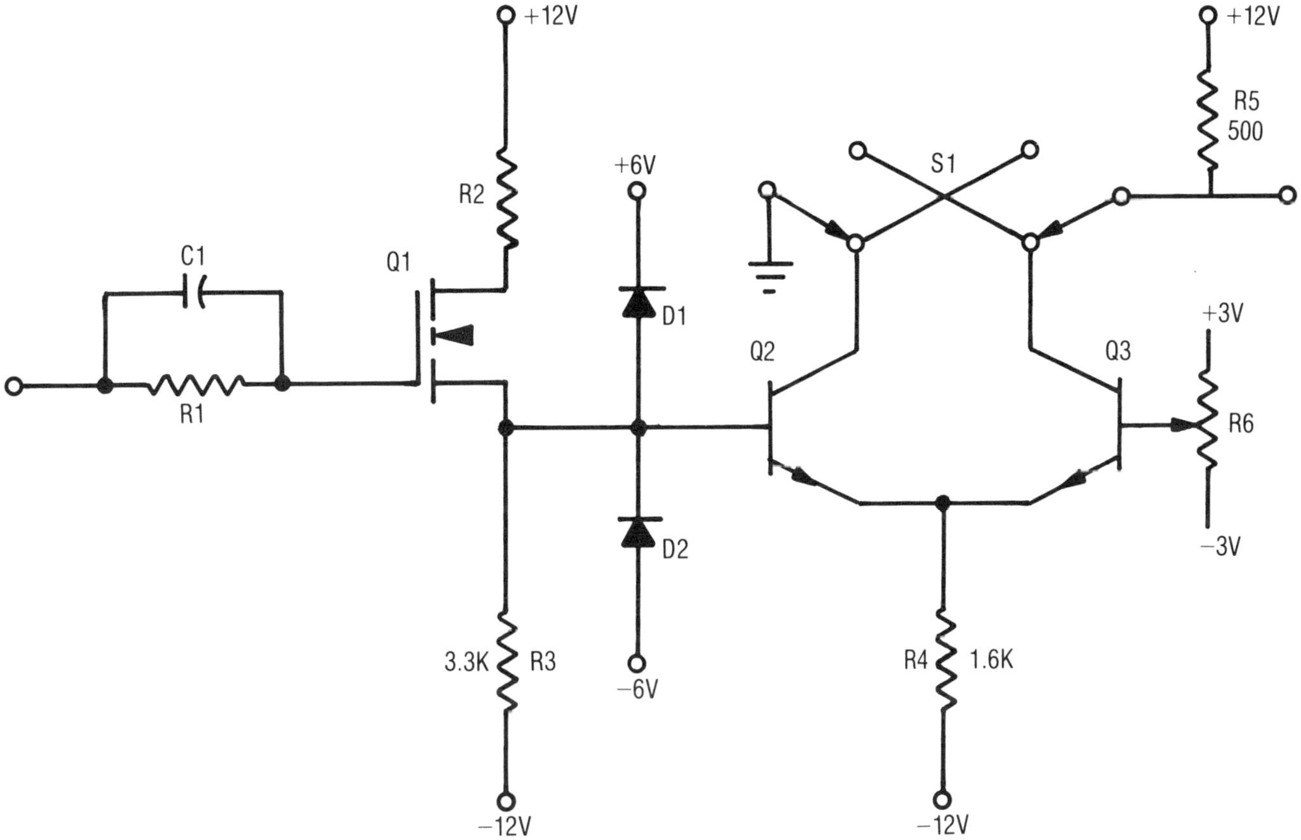

Figure 5.5 Transistor trigger input amplifier

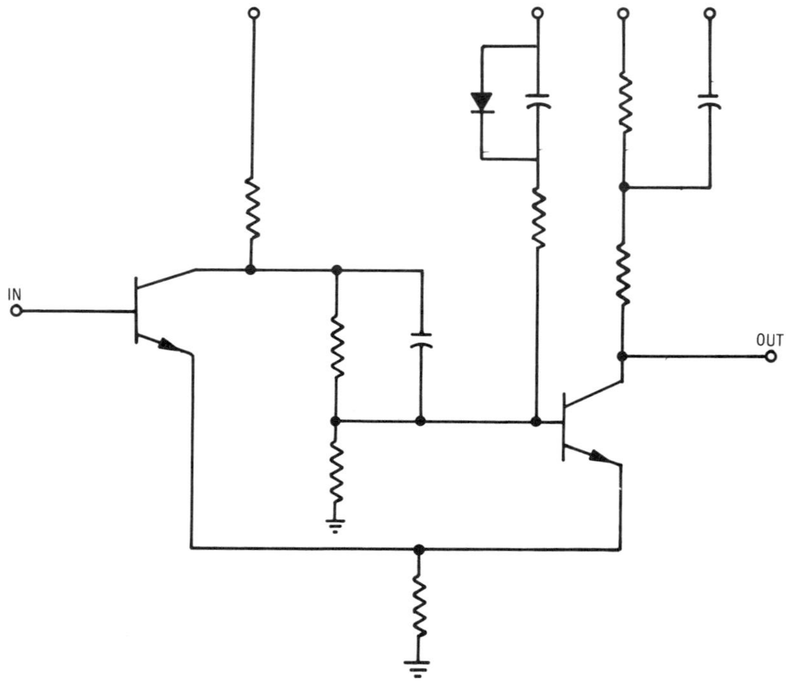

Figure 5.6 Schmitt trigger with transistors

Previously, we mentioned the "auto" triggering mode. If trigger-circuit controls are not set properly, no sweep will be generated, so no trace will appear. This is often a problem at higher frequencies. Also, you must have something on the screen to set the controls correctly!

You can, of course, adjust for a free-running sweep, as we saw in the last chapter. Or, you can use the "auto" position of the level control or trigger mode switch (it may be on either).

When placed in this position, the control connects a resistor between the grids of V1 and V2 in Figure 5.4, and sets the level control voltage at zero. This causes the pulse generator to free-run at a rate determined by the time constant of the resistor and the input capacitor (this mode is used only with a-c coupling). This is a fairly low frequency. A trace will therefore appear even in the absence of a signal. However, when a signal is present it will control the pulse generator operation as long as it has a higher frequency.

Finally, before the output pulses of the pulse generator are applied to the input of the sweep-gating multivibrator (described in the previous chapter) they are passed through a differentiator circuit that converts them to sharp spikes. This is simple RC arrangement as in Figure 5.7. Since the differentiator produces both positive and negative spikes, a diode is added to ground the unwanted positive ones.

Horizontal Amplifier

In most scopes the main task of the horizontal amplifier is to convert the single-ended sawtooth output from the sweep generator into a push-pull signal suitable for driving the horizontal deflection plates of the CRT, as shown in Figure 2.1 in Chapter 2.

The amplifier is similar to many d.c.-coupled audio amplifiers, as you can see from the basic circuit in Figure 5.8. In this example, the sawtooth signal is applied across a voltage divider R1 and R2. The values of these resistors are chosen so that the voltage at the S1 terminal "X1" is one-tenth of that at "X10." The switch S1 is used to select either of these input signals for coupling via S2 and R3 to the grid of V1A.

Sweep Magnifier

R1, R2 and S1 constitute a *sweep magnifier*. It is provided in many scopes to help make more accurate measurements between points on the horizontal axis of the display that are close together.

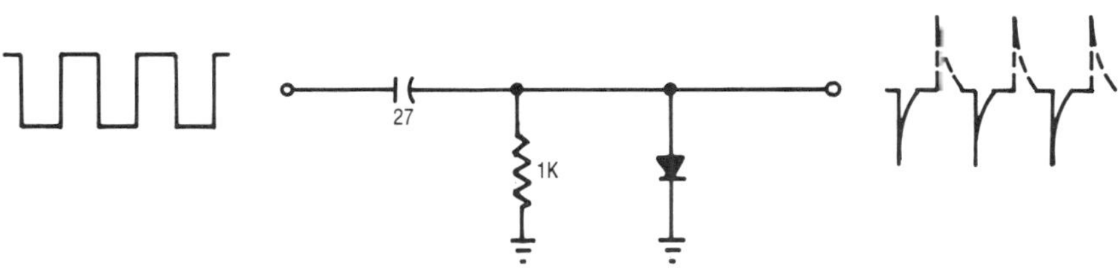

Figure 5.7 Action of differentiator

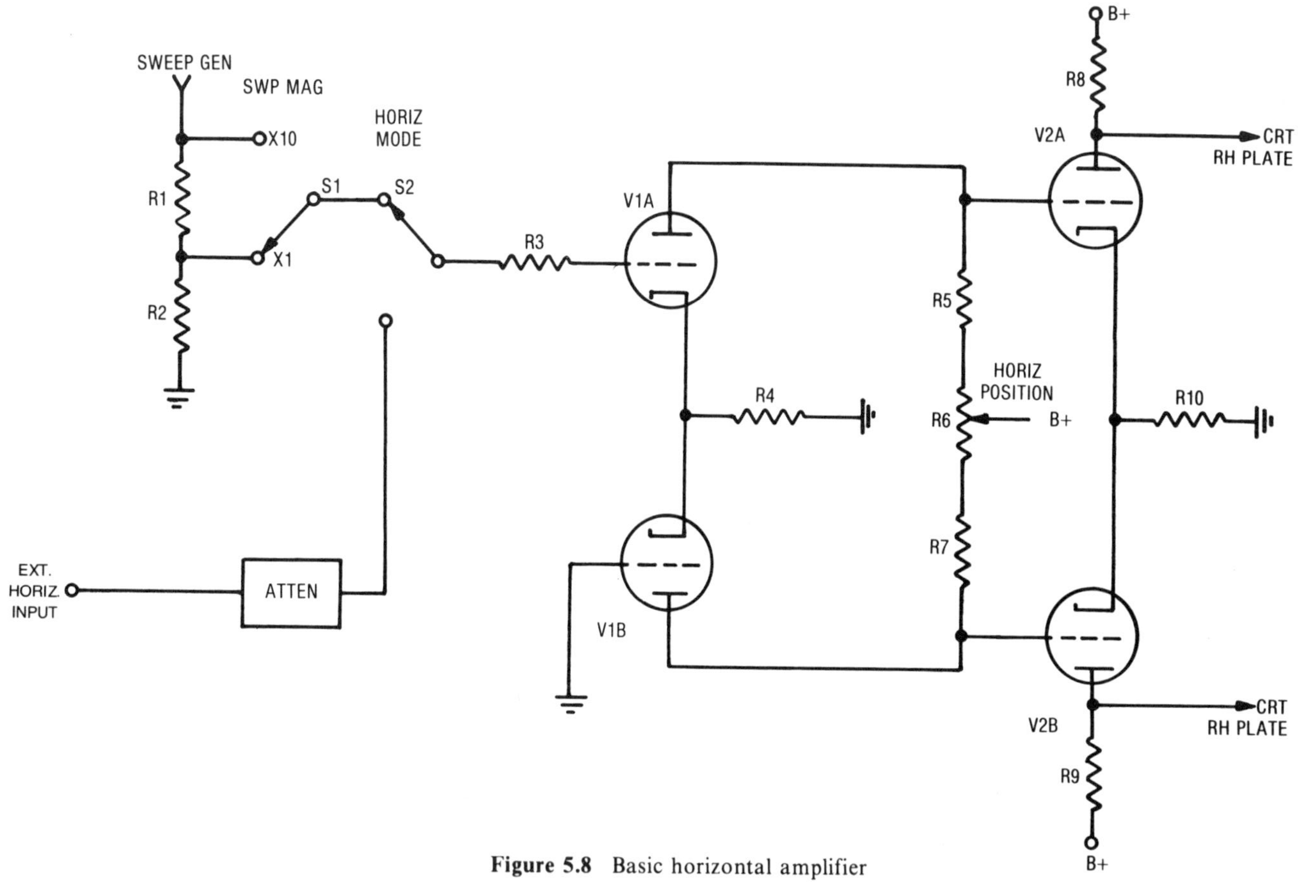

Figure 5.8 Basic horizontal amplifier

The X1 position is the normal setting of S1, and when the switch is in this position the amplitude of the sawtooth signal applied to V1's grid is of the proper size to give a sweep, after it has been amplified, that is the exact length of the ten graticule divisions across the scope screen. When S1 is reset to its X10 position the amplitude of the signal applied to V1A's grid is ten times as great, so the sweep would now cover 100 divisions of the screen, if it had them. Two points on the trace that were only one centimeter apart are now ten centimeters apart, so resolution is greatly improved. Since only ten divisions of the theoretical 100 divisions can be viewed on the screen, the display would have to be moved laterally to get the section you want where you can see it. This is done with the *horizontal position control,* which we'll come to later.

Another advantage of the sweep magnifier is that it extends the sweep speed. Suppose the fastest sweep rate available on the time-base control is 1 microsecond per division. The sweep magnifier stretches this out to 1 microsecond per ten divisions, which is equivalent to 0.1 microsecond per division.

While X10 is a common magnification factor, you may come across other factors from X2 to X100, and in some cases several magnification factors are available on the switch, or a variable sweep magnification control.

Horizontal Mode

S2 connects the amplifier input either to S1, so that the sawtooth signal from the sweep generator is applied to the grid of V1A, or to the external horizontal input. S2 may be a separate control or it may be a setting on the time-base timing switch.

Obviously, there's no sweep when S2 is in its external horizontal input position. Instead, the amplifier amplifies whatever signal is applied to the external horizontal input connector on the scope front panel. This will appear as a voltage on the horizontal deflection plates in the same way that a signal appears on the vertical deflection plates, except that its amplitude is horizontal instead of vertical. If the same sine-wave signal is connected to both the vertical and horizontal inputs and their respective attenuators adjusted for equal deflections, the display will consist of a combination of the vertical and horizontal signals, as shown in Figure 5.9(a). The voltage on

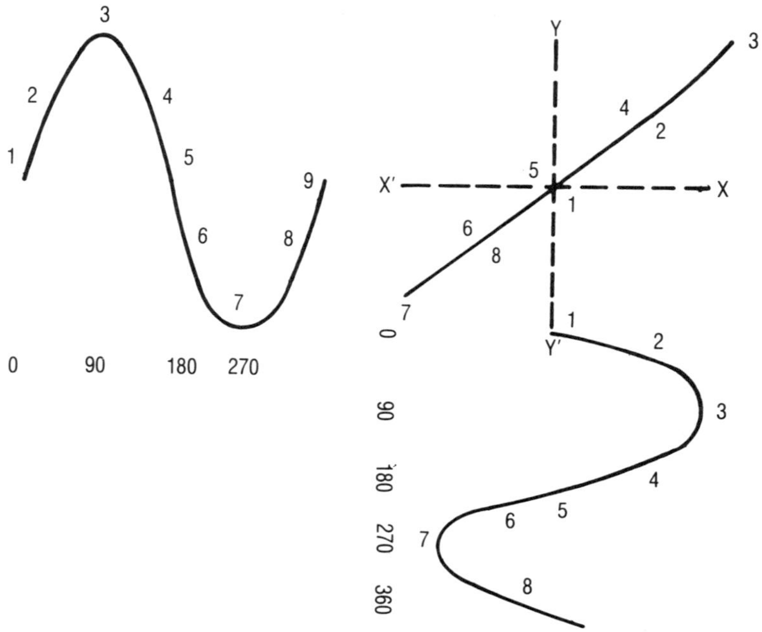

Figure 5.9 (a) X-Y display of two signals of same frequency phase. One signal is applied to the Y input, the other to the X input. The resulting display is a sloping line.

the vertical deflection plates rises and falls in accordance with the vertical input, and if there were no voltage on the horizontal plates the electron beam would trace out the line Y′-Y. Similarly, if there were no voltage on the vertical deflection plates, the horizontal signal would give us the line X′-X. When both pairs of plates are used the resultant display is a slanting line as shown.

However, suppose the signal at the horizontal input is phase-shifted with respect to the signal at the vertical input. Figure 5.9(b) shows what happens when the horizontal signal lags the vertical signal by 45 degrees. The slanting straight line has become an ellipse. With a 90-degree phase shift, the ellipse becomes a circle, as in Figure 5.9(c).

These displays are called *Lissajous figures*. Those indicating phase shift are important to electronics technicians. They are often used to adjust a device for zero phase shift by setting the phase

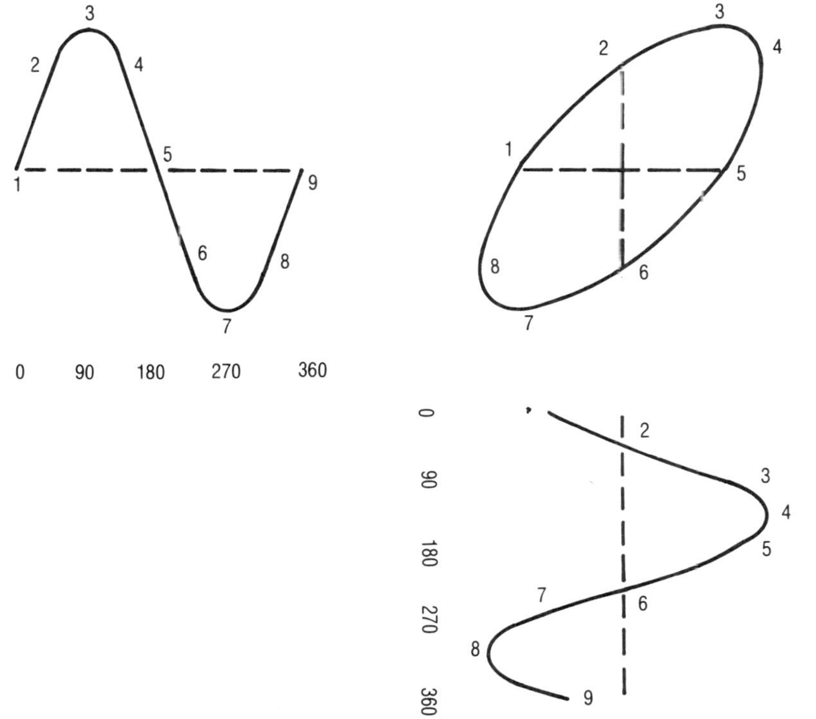

Figure 5.9 (b) X-Y display of two signals with a 45° phase shift with respect to each other. The signal at the left leads the signal at the right. The resulting display is an ellipse.

adjustment to the position where the ellipse becomes a straight line (0 degrees).

Others may be formed by combining signals of different frequencies as shown in Figure 5.10. A common use of this is in measuring modulation percentage, as we shall see in a later chapter.

If you need really accurate results, as in adjusting synchros, you should use a scope specially designed for this. Such a scope has identical vertical and horizontal amplifiers to avoid the introduction of spurious phase shift by the scope itself. It also does not have a delay line, since this would be another cause of false phase shift. It follows from this that the *same* scope cannot be a high-efficiency X-Y scope and a highly efficient high-frequency scope. Some compromises are necessary in general-purpose scopes, and these usually

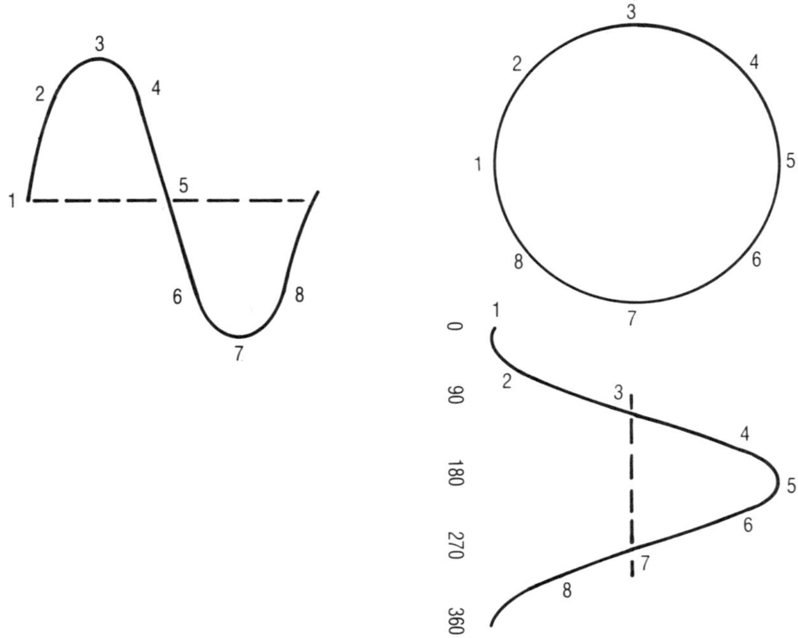

Figure 5.9 (c) X-Y display of two signals with a 90° phase shift. The signal at the left leads the signal at the right. The resulting display is a circle.

provide for better high-frequency response than for X-Y performance.

Paraphase Amplifier

Returning to Figure 5.8, V1A and V1B are the active devices of a paraphase amplifier. This circuit changes the single-ended sawtooth signal appearing on the grid of V1A into push-pull outputs that drive V2A and V2B. As the voltage rises on V1A's grid the current through the tube increases. The d.c. electron path for this current is from ground via R4, V1A, R5 and R6 to B+, and as the current increases the voltage drop across each of these components rises also.

Since one end of R4 is connected to ground, the increase in the voltage drop across this resistance shows up as a higher positive potential at the end of R4 connected to V1A's cathode. However,

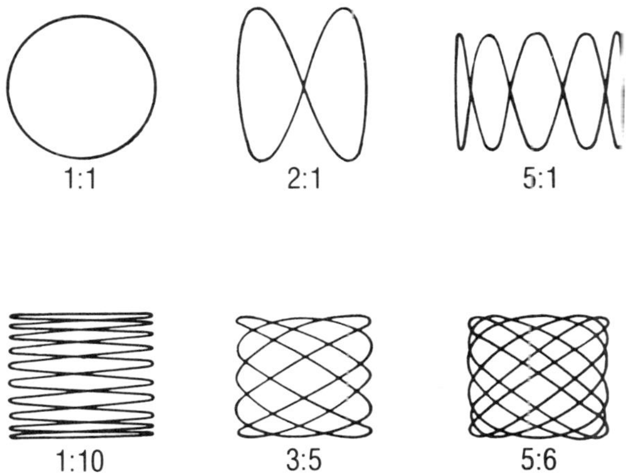

Figure 5.10 Lissajous patterns for various frequency ratios

this end is connected to V1B's cathode as well, so the same voltage appears here. As a result, V1B's cathode becomes more positive with respect to its grid, which remains at zero volts whatever happens, because it is tied to ground.

You could just as well say the V1B's grid becomes more negative with respect to its cathode as far as the effect is concerned, which is that the grid offers an increasing opposition to electron flow from cathode to plate as the cathode goes more positive. Consequently, fewer and fewer electrons make it to the plate. This diminishing current has exactly the opposite effect to that through V1A, because the voltage drops decrease across each component in its path.

As the potential drop increases across R5 and the portion of R6 between the end connected to R5 and the sliding contact, the voltage at the end of R5 connected to V1A's plate decreases, since B+ does not change. This results in a decreasing voltage appearing on V2A's grid. At the same time the voltage drop across R7 and the other portion of R6 is decreasing, so a rising voltage appears on V2B's grid.

Output Amplifier

The outputs of V2A and V2B are, of course, 180 degrees opposite to their inputs, so we get a rising positive voltage on V2A's plate that is coupled to the right-hand horizontal deflection plate of the CRT. At the same time, the output of V2B is decreasing, and this appears on the CRT's left-hand plate.

The CRT beam will only be centered horizontally when both deflection plates are at the same potential, which will be the case when V2A and V2B have equal outputs. Because of small variations in component performance these may not be identical, so the horizontal position control is provided to equalize the grid voltages of the output tubes to bring this about. Another problem that may arise is that the display center point shifts from the middle of the screen when the SWP MAG switch is repositioned from X1 to X10. An additional internal adjustment (not shown in Figure 5.8) is often provided to correct this during calibration.

Solid-State Circuits

Most modern scopes have solid-state horizontal amplifiers. Figure 5.11 is an example. The sweep signal is applied to the base of Q1 and the external horizontal input to the base of Q2. When the sweep is being used the base of Q2 is grounded, and when the external horizontal input is selected the base of Q1 is grounded. As there is no common emitter resistor these transistors do not operate as a paraphase amplifier. Their output is always single-ended, and is applied either to the base of Q3 or to that or Q4.

These transistors form a paraphase amplifier that acts in a similar way to the previous example. They drive the output stages consisting of Q5, Q6, Q7 and Q8.

In the emitter circuit of Q3 and Q4 is the MAG SWP switch. In its 10X position this switch shunts R16 and R17 with R19 and R20. This increases the gain of this stage approximately ten times. R17 and R20 are adjustments to set the sweep to exactly ten divisions in

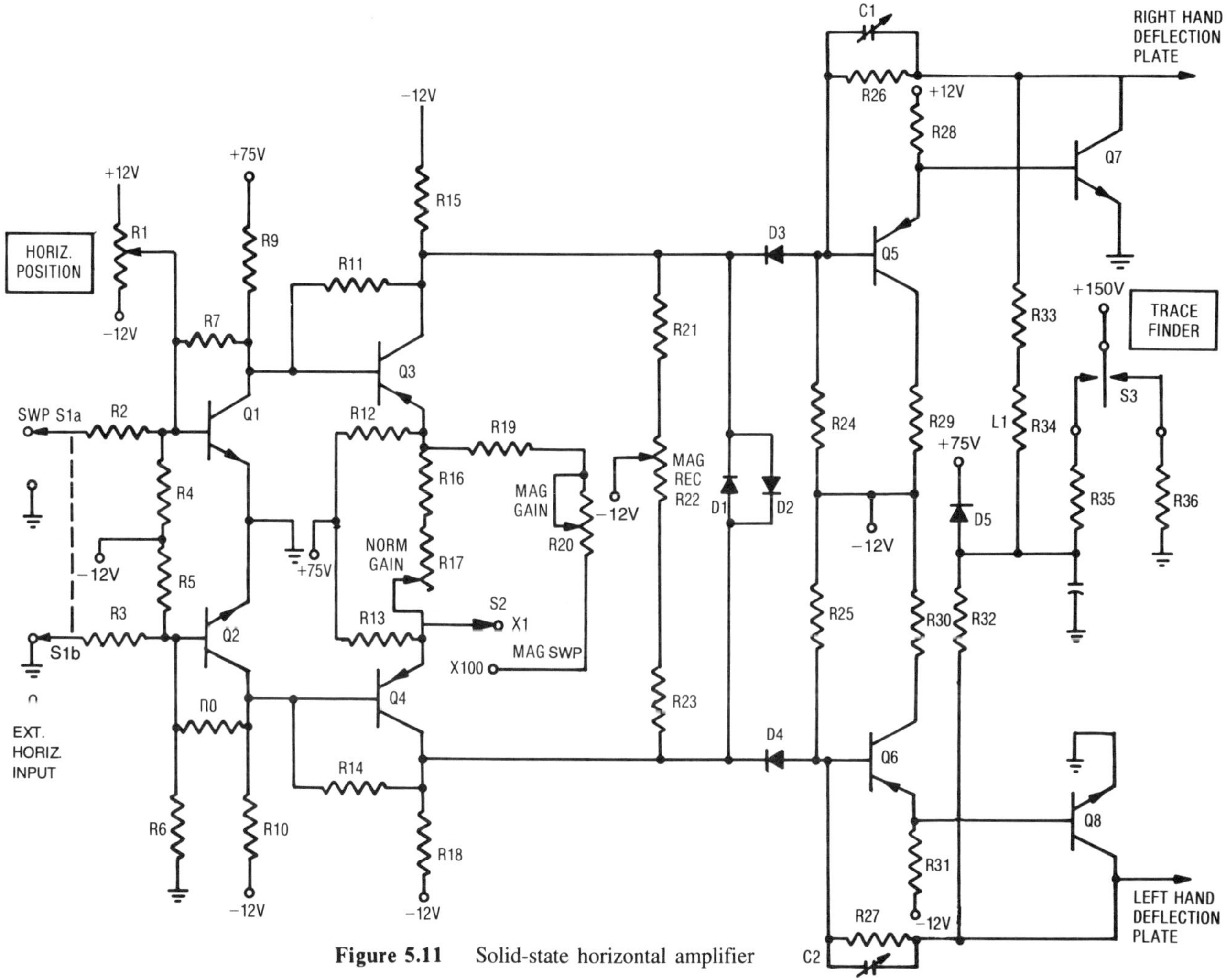

Figure 5.11 Solid-state horizontal amplifier

either position of the MAG SWP switch. The MAG REG adjustment R22 is set to prevent any sideways shift of the trace when switching between X1 and X10.

The horizontal position control works by varying the bias on Q1's base. This causes opposite shifts in the d.c. level on the two sides of the push-pull amplifier so that the trace is displaced sideways on the screen.

One of the main differences between vacuum-tube amplifiers and their junction-transistor counterparts is the low input impedance of the latter. In a vacuum tube, variations in the plate current do not change the tube's loading effect on the previous circuit. This is not the case, however, with a transistor which can go into saturation or cutoff with adverse effect on the sweep linearity.

In the circuit of Figure 5.11 this is prevented by diodes D1 through D4. If we are using the magnified sweep, Q7 would saturate as the sweep signal tried to drive the beam off screen to the left. Saturation causes a delay in the collector's response to the signal on the base. This would result in an inaccurate sweep. Therefore, as the collector of Q3 drops to about 0.5 volt, D3 becomes back-biased and disconnects. This prevents any further change at the bases of Q5 and Q7, so that Q7 does not saturate. When the sweep ramp rises to a level that turns D1 on again, Q7 follows suit without any delay due to saturation. The same action occurs with Q4, D4, Q6 and Q8.

When D3 or D4 disconnects, Q3 or Q4 loses its low-impedance load, and this would tend to go into saturation. D1 and D2 prevent this, however, since one or the other turns on to maintain the required low impedance.

Trace Finder

The push-button switch S3 disconnects the collector voltage of Q7 and Q8, so that if the beam were positioned off the screen it would come back on as we have removed the deflection voltages. The switch connects the supply voltage to ground via R36 so the load on the power supply is not decreased enough to cause the supply to other circuits to increase, affecting their operation.

Intensity Modulation

Many scopes have on their rear panels a pair of terminals labeled "Z axis input." One is a ground terminal, and the two are usually connected by a ground strap. This should always be in place, except when you are using this input. When the strap is disconnected, a signal applied to the Z-axis input terminal is capacitively connected to the CRT grid or cathode. If the polarity of the signal increases the negative bias of the grid with respect to the cathode, the intensity of the electron beam is decreased, or even blanked altogether, so that the trace on the screen is darkened or extinguished. On the other hand, if the negative bias is decreased by the Z-axis input signal, the trace is brightened. A square-wave signal of the proper amplitude will chop the trace into a dashed line, the length of the dashes depending upon its frequency.

This is called Z-axis modulation or intensity modulation, and may be used for marking time intervals on the trace when using an uncalibrated sweep or operating in the X-Y mode.

Single Sweep

When the signal to be displayed is not repetitive or varies in amplitude, shape or time, a conventional display may be impossible because of instability. To avoid this, some scopes have a single sweep mode. This mode can also be used to photograph a non-repetitive signal.

To use the single sweep mode, first make sure the trigger circuit will respond to the event to be displayed. Set the triggering mode to auto or normal, and obtain the best possible display of the unstable signal. Or if you can't do this, set the trigger circuit to trigger on a signal that is approximately the same amplitude and frequency as the random signal. Then set the sweep mode to single sweep and press the reset button. When this is pushed the next trigger pulse causes the sweep to run once. It is then locked out until you press the reset button again. Scopes with this feature usually have a light labeled "reset" that lights to tell you that the sweep is "armed" or ready to go. It goes out after the sweep, and remains off until you press the reset button again.

6

Using Waveforms

In the catastrophic type of problem, where something has broken down so that the equipment does not perform at all, signal tracing (see the following chapter) will identify the defective stage by telling you where the signal disappears. If, for instance, in Figure 6.1 it is present at C, but absent at D, it is obvious something is wrong in the modification circuits. If there is no output at all, you'd start looking for an open, a short, or a bad tube or transistor.

However, in the type of problem where the complaint is improper performance, identifying the defective stage is not always so simple. Distorted sound from a radio, for example, could not be caused by a dead battery, open resistor or shorted transistor because any of these defects would kill the set. The cause of the garbled sound must be something else that allows the radio to play, but with impaired performance. Even so, a *major* change in equipment operation, such as weak operation, squealing, motor-boating and so on, in our radio will also be fairly easy to track down, but a *minor* change that causes a slight deterioration in overall operation requires a more refined technique. The two most effective for checking amplifiers are *sine-wave* and *square-wave analysis*.

For these tests you need your scope and a signal generator that produces sine-waves and square-waves. The basic test equipment set-up is shown in Figure 6.2. The unit under test must be terminated with a load that simulates whatever it works into. In the case of a hi-fi amplifier you would use a 25- or 50-watt non-inductive resistor with a resistance equal to the impedance of the loudspeaker normally driven by the amplifier.

All the equipment controls are set for normal operation to begin with, although you may need to alter them later. The signal

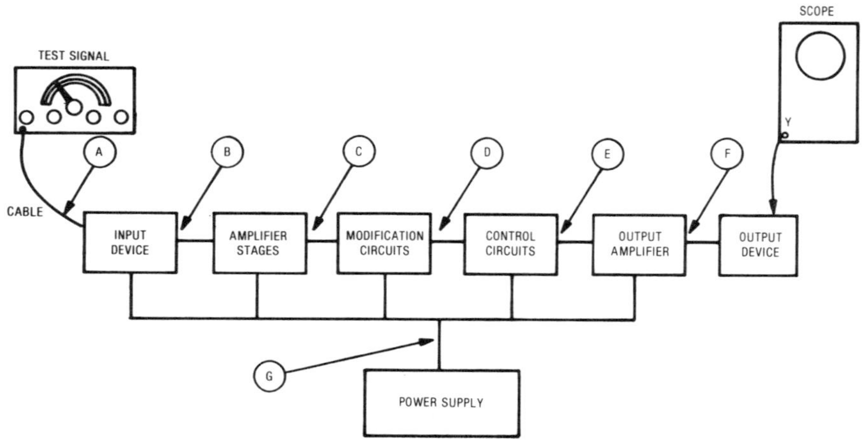

Figure 6.1 Block diagram of typical sections in electronic equipment

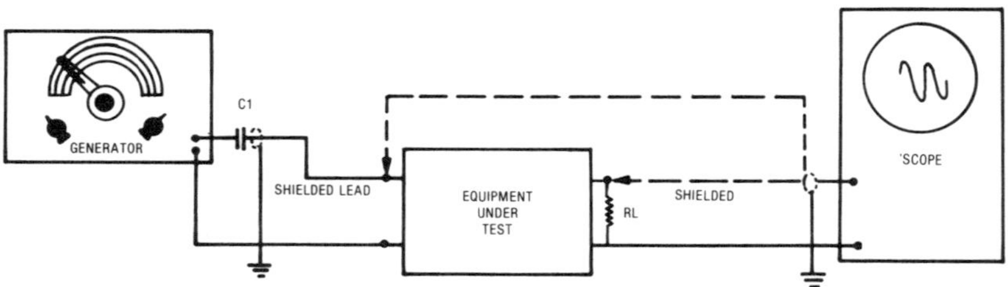

Figure 6.2 Basic equipment set up for sine-wave or square-wave performance tests

generator is connected to the input through a shielded cable with an isolating capacitor as shown in Figure 6.2. This capacitor should have a fairly high value, preferably from five to ten times that of any interstage coupling capacitors in the unit being tested. The signal generator's output should initially be set to give an input signal with about the same amplitude as the equipment's normal input.

You should first connect the vertical input of your scope to the input of the unit to see what kind of input it is receiving from the signal generator. Any changes in subsequent stages must, therefore, be caused by the circuits through which the signal passes. The schematic diagram and your own knowledge of circuit performance will lead you to expect certain waveforms at the points probed.

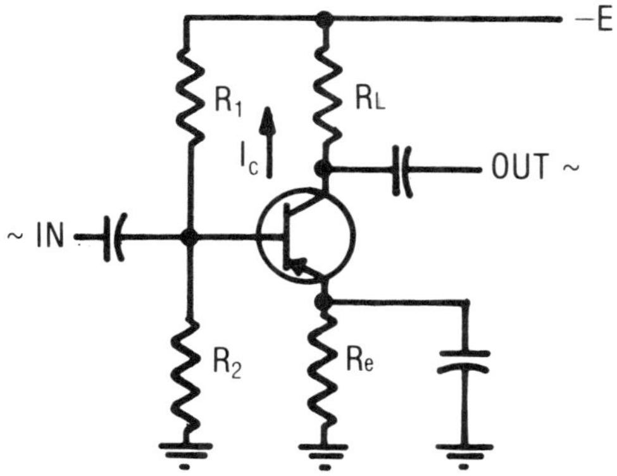

Figure 6.3 Single-stage transistor amplifier

When you find a deviation from normal you will then investigate the circuit between that point and the point before it where the waveform was correct.

Sine-wave tests are performed to check stage gain and frequency response. For example, consider the circuit in Figure 6.3. This is a single-stage transistor amplifier. If the input signal is normal and the output is also normal (no distortion), the circuit operation is also normal. We should be able to measure the amplitude of the input signal and divide it into the amplitude of the output signal and obtain a figure for the stage gain that is close to the theoretical figure.

Let us suppose, for instance that the amplitude of the input signal is 60 millivolts peak-to-peak. With the vertical amplifier volts/division control set at 10 mV/div, the sine-wave will occupy six divisions vertically, as shown in Figure 6.4(a). When we transfer the scope probe to the output the display gets much too large for the screen, so we readjust the volts/division control until the waveform fits on the screen. Then we use the vertical position control to place the negative peaks on the same baseline as we used for the input signal.

Suppose the amplitude of the signal is now 5.2 divisions with the volts/division control set at 1 V/div, giving us a display as in Figure 6.4(b). This means that the amplitude of the output signal is 5.2 volts. Dividing this by 60 millivolts gives a stage gain of—

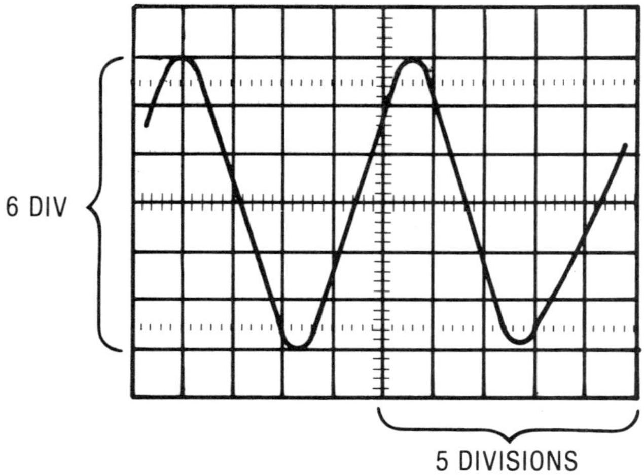

Figure 6.4 Measurement of stage gain. **(a)** The input signal has an amplitude of 60 mV (6 divisions with the volts/division control set at 10 mV/division)

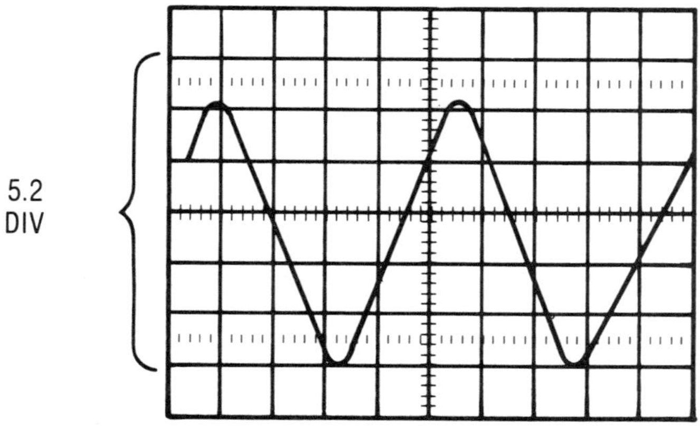

Figure 6.4 (b) The output signal has an amplitude of 5.2 V (5.2 divisions with the volts/division control set at 1 V/division)

$$\frac{5.2}{0.06} = 87$$

so we know our stage is working.

However, even if the stage gain is satisfactory, there may be other problems. Suppose our output signal is distorted in one of the ways illustrated in Figure 6.5. We may find we have a problem of a different kind, as in Table I on page 128.

If You Don't Know the Frequency of the Displayed Signal

As we've mentioned already, the time interval between successive cycles is easily measured on the graticule in the horizontal direction. In Figure 6.4, for instance, the distance between the point

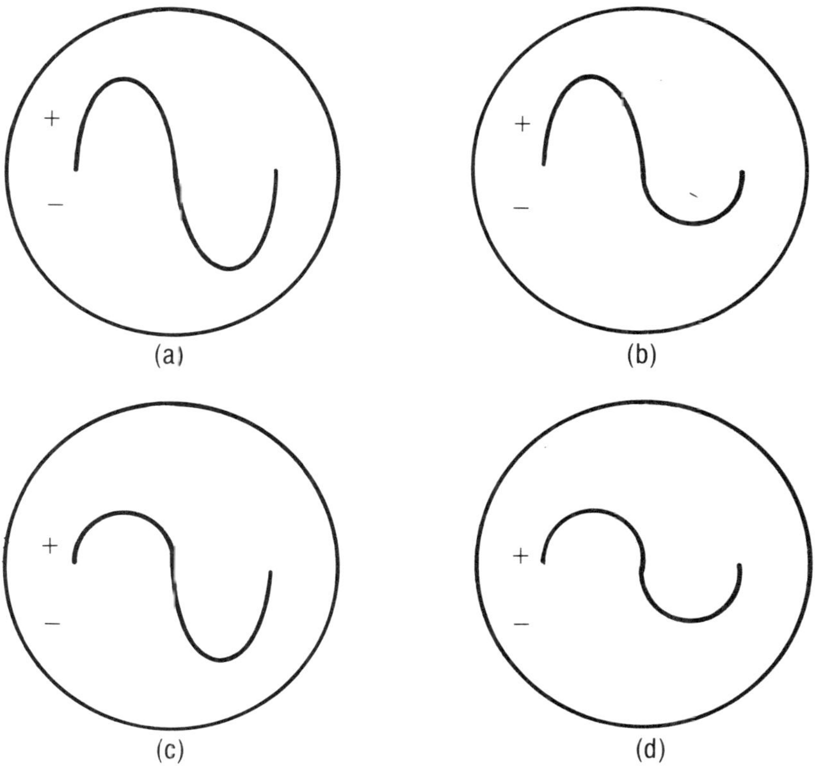

Figure 6.5 Various types of sine-wave distortion . . . (a) Normal waveform, (b) Flattened negative peak, (c) Flattened positive epak, and (d) Both peaks flattened or clipped

TRANSISTOR TYPE	SINE-WAVE DISTORTION	SEE FIGURE	PROBABLE CAUSE
PNP	None	6.5(a)	Circuit operation normal
	Negative peaks clipped	6.5(b)	Low bias; battery voltage low; weak transistor; high-value load resistor.
	Positive peaks clipped	6.5(c)	High bias; weak or high-resistance battery; leaky transistor; leaky coupling capacitor; low-value load resistor.
	Both peaks clipped	6.5(d)	Overdrive; weak battery; weak transistor.
NPN	None	6.5(a)	Circuit operation normal
	Negative peaks clipped	6.5(b)	High bias; weak or high-resistance battery; leaky transistor; leaky coupling capacitor; low-value load resistor.
	Positive peaks clipped	6.5(c)	Low bias; battery voltage low; weak transistor; high-value load resistor.
	Both peaks clipped	6.5(d)	Overdrive; weak battery; weak transistor.

Table I. Sine-wave distortion reference chart

where the rising portion of the first cycle crosses the horizontal center line and the point where the rising portion of the second cycle crosses it is five divisions. If the time/division control had been set to 1 millisecond (per division), the time for one cycle of this signal would be 5ms. Its frequency would, therefore, be $1/5 \times 10^{-3}$ or 200 hertz.

This little calculation is really easy to do on any pocket calculator. The most important thing is to remember that the time for one cycle must be entered in *seconds*. Five milliseconds would therefore be entered as .005. Five microseconds would be entered as .000005. (However, if your calculator has scientific notation, you would enter them as 5×10^{-3} or 5×10^{-6}.)

If your calculator gives reciprocals ($^1/x$) directly, you have only to enter the time in seconds and then depress the reciprocal key to obtain the frequency in hertz. The reciprocal of 5×10^{-3} will appear as 200, that of 5×10^{-6} as 200000 (or 200 kHz). However, if your calculator does not have a reciprocal key, first enter 1, then ÷, followed by the time. This is dividing the time into 1, which is what a reciprocal is anyway.

Checking the Frequency Response of an Amplifier

To check the response of the circuit to different frequencies, apply one frequency at a time to the input. Connect the scope to the input also, and adjust the signal generator's output to give a display on the scope that fits exactly between two graticule lines, as in the example in Figure 6.4(a).

Then transfer the scope probe to the output, as shown in Figure 6.2, and observe the display. Write down the amplitude and frequency.

Now reconnect the scope to the input, and adjust the signal generator for the next frequency. If necessary, readjust the output attenuator to maintain the same input signal level.

Move the probe back to the output, and write down the new signal amplitude and frequency, as shown on the scope. Repeat this for the range of frequencies you wish to cover.

This is a good method for determining the effectiveness of some controls in adjusting an amplifier's frequency response characteristics, or for any other procedure where you want to obtain precise frequency response measurements. However, a more rapid check of overall frequency response can be made using square waves instead of sine waves.

The test setup is the same as for sine-wave testing, except that the signal source is a square-wave generator. It is, however, important to keep the amplitude of the input signal as low as possible, consistent with getting an adequate response, because it is possible to overload the circuit without realizing it. With a sine wave this shows up in flattened peaks, but a square wave is already flat-topped! An overloaded circuit will give you false results, and could even be damaged.

A square wave may be distorted in various ways, as shown in Figure 6.6. The interpretations of these distortions are given in Tables II, III and IV.

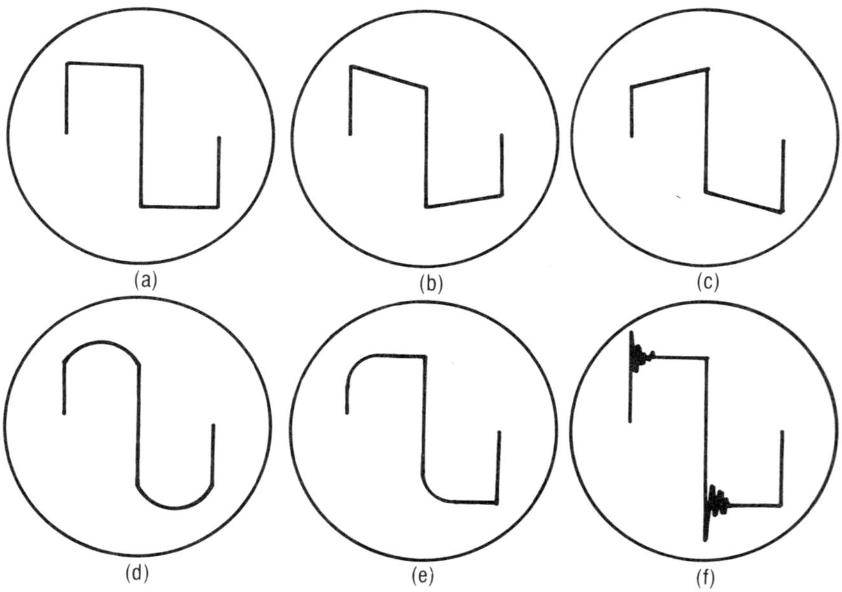

Figure 6.6 Various types of square-wave distortion . . . (a) Normal waveform, (b) and (c) Slanting (or tilted) top, (d) Rounded top, (e) Rounded leading edge, and (f) "Ringing" or damped oscillation

SQUARE-WAVE INPUT	OUTPUT WAVEFORM	FIGURE NO.	DIAGNOSIS
Low frequency (50-100 Hz)	Like input	6-6 (a)	Circuit operation normal.
	Slanting down	6-6 (b)	Poor low frequency response—partially open coupling or emitter by-pass capacitor.
	Sharp pulse	—	Extremely poor low frequency response—open coupling capacitor or shorted primary in interstage transformer.
	Slanting up	6-6 (c)	Accentuated low frequency response—partially open by-pass capacitor or defective decoupling filter.
	Rounded top	6-6 (d)	Accentuated response at fundamental frequency of square wave—defective interstage transformer, changed value coupling or by-pass capacitors, defect in degenerative feedback network.
	Rounded leading edge	6-6 (e)	Reduced upper frequency response—excessive distributed capacities, defective interstage transformer.
	Damped oscillation	6-6 (f)	Peak in middle range of frequencies—changed value by-pass capacitor, defective coupling transformer, inadequate shielding, poor lead dress, defect in feedback network, defective by-pass or decoupling capacitors.

Table II. Square-wave distortion reference chart—low frequencies

SQUARE-WAVE INPUT	OUTPUT WAVEFORM	FIGURE NO.	DIAGNOSIS
Medium frequency (500-2500 Hz)	Like input	6-6 (a)	Circuit operation normal.
	Slanting down	6-6 (b)	Phase shift, poor low-frequency response—partially open coupling or emitter by-pass capacitor.
	Sharp pulse	—	Extremely poor low and middle-frequency response; open coupling capacitor or shorted primary in interstage transformer.
	Slanting up	6-6 (c)	Accentuated response at middle frequencies—partially open by-pass or filter capacitor, defect in feedback network.
	Rounded top	6-6 (d)	Accentuated response at fundamental frequency of test square-wave signal—defective interstage transformer or partially open by-pass capacitor, defect in feedback network, poor lead dress.
	Rounded leading edge	6-6 (e)	Reduced upper to high-frequency response—excessive distributed capacities, defective interstage transformer.
	Damped oscillation	6-6 (f)	Peak at high frequencies—changed value by-pass capacitor, defective coupling transformer, poor shielding, improper lead dress, defect in feedback network, open load resistor (in video circuits).

Table III. Square-wave distortion reference chart—medium frequencies

SQUARE-WAVE INPUT	OUTPUT WAVEFORM	FIGURE NO.	DIAGNOSIS
High frequency (5kHz—20 kHz)	Like input	6-6 (a)	Circuit operation normal—high-frequency response reasonably flat to 10 or 20 times input square-wave frequency.
	Slanting down	6-6 (b)	Phase shift, poor medium-frequency response—partially open coupling or emitter capacitor.
	Sharp pulse	—	Extremely poor medium-frequency response—open transistor or interstage transformer.
	Slanting up	6-6 (c)	Accentuated response at high frequencies, causing phase shift—defect in frequency compensation network, changed-value loading resistors (in video amplifiers).
	Rounded top	6-6 (d)	Accentuated response at fundamental frequency of test signal—should show up as "damped oscillation" when lower-frequency test signal used. Open by-pass or decoupling capacitor, poor lead dress, defective shielding, defect in feedback network.
	Rounded leading edge	6-6 (e)	Poor high-frequency response, with "rounding" increasing as high-frequency response falls off—excessive distributed capacities, increased value load resistor, open peaking coil (in video circuits).
	Damped oscillation	6-6 (f)	Peak in response at frequency beyond normal (desired) range—open by-pass or decoupling capacitors, poor lead dress, defective shielding, open load resistor (in video circuits), wrong value or misadjusted peaking coil, improperly loaded interstage transformers.

Table IV. Square-wave distortion reference chart—high frequencies

Evaluating Modulation

In the case of a transmitter, you may need to measure the *modulation index* of the signal at the output. There are three ways in which you can do this with a scope, using the transmitter's output carrier modulated with a sine-wave of audio frequency.

In the *trapezoidal pattern* method the scope is connected to the transmitter as shown in Figure 6.7. This applies the modulated output signal to the horizontal input. The sweep is disabled, so in the absence of a modulating signal the unmodulated carrier deflects the electron beam up and down only, as in Figure 6.8(a).

As soon as a modulating signal is applied to the horizontal input the electron beam is deflected back and forth sideways as the signal goes negative and positive. As the amplitude of the carrier is also rising and falling in accordance with the modulations imposed upon it, the resulting display takes the form of a trapezoid, as in Figure 6.8(b).

In 100-percent modulation, the amplitude of the modulated signal varies between zero and twice the amplitude of the unmodulated carrier, so the display becomes a triangle with the apex to the

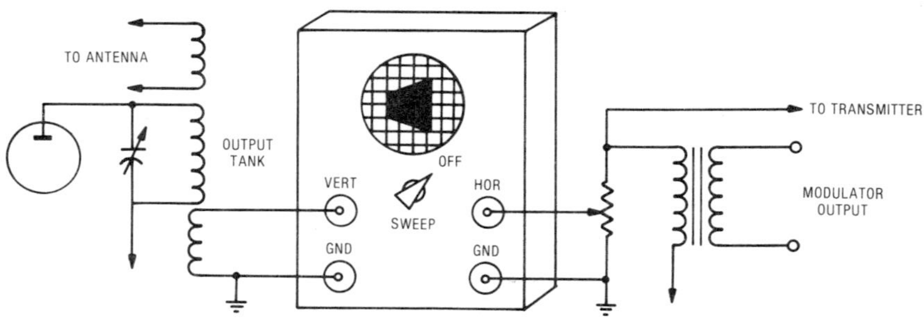

Figure 6.7 Connections for obtaining trapezoidal patterns

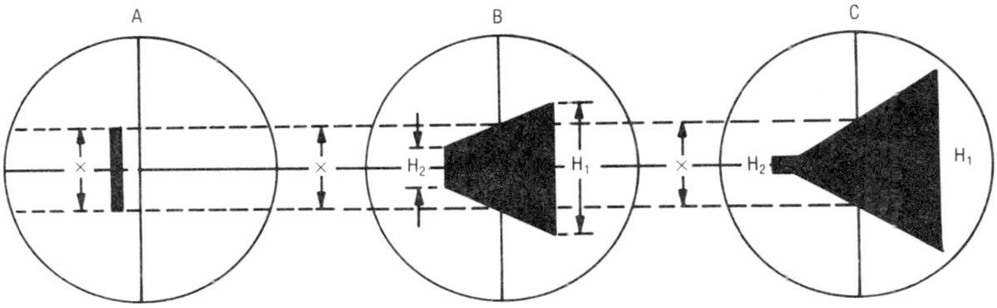

Figure 6.8 Typical trapezoidal patterns

left. This is the zero, while the base to the right is produced by the peak-to-peak value. The modulation index is given by:

$$\frac{H_1 - H_2}{H_1 + H_2} \times 100$$

which in this case is therefore 100 percent.

In the second display, (b), H_1 is six divisions while H_2 is only two. This translates to:

$$\frac{(6 - 2)}{(6 + 2)} \times 100 = \frac{4}{8} \times 100 = 50 \text{ percent modulation.}$$

In the third display, (c), H_2 is elongated and brightened. This indicates over-modulation, which produces distortion in the signal.

In Figure 6.9 a different hookup is illustrated. Both horizontal and vertical inputs are connected to the transmitter's output, but a 50-kilohm resistor is connected in series with the horizontal input. This results in a phase shift between the vertical and horizontal input signals, because the resistor combines with the scope's horizontal input capacitance to produce an RC time constant. If the latter is 30 pF, a typical value, the time constant will have a value of:

$$5 \times 10^4 \times 3 \times 10^{-11} = 1.5 \times 10^{-6} \text{ second, or}$$

1.5 microseconds.

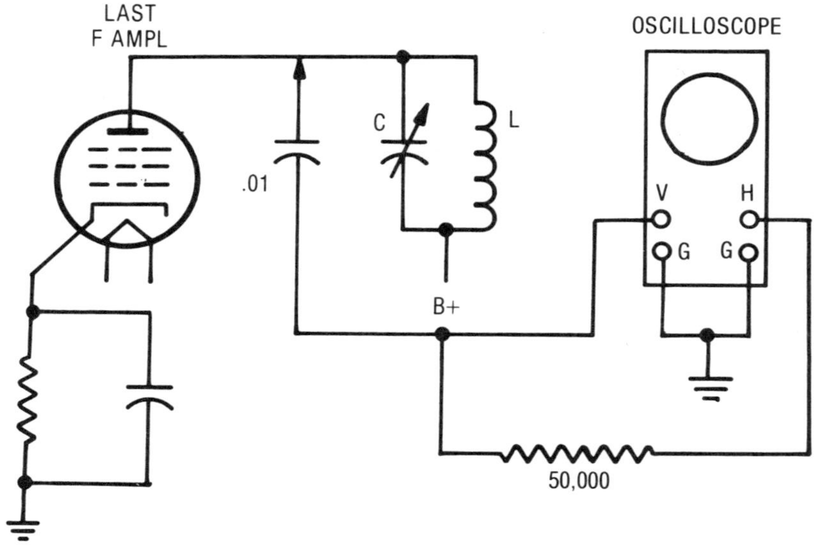

Figure 6.9 Connections for Lissajous patterns

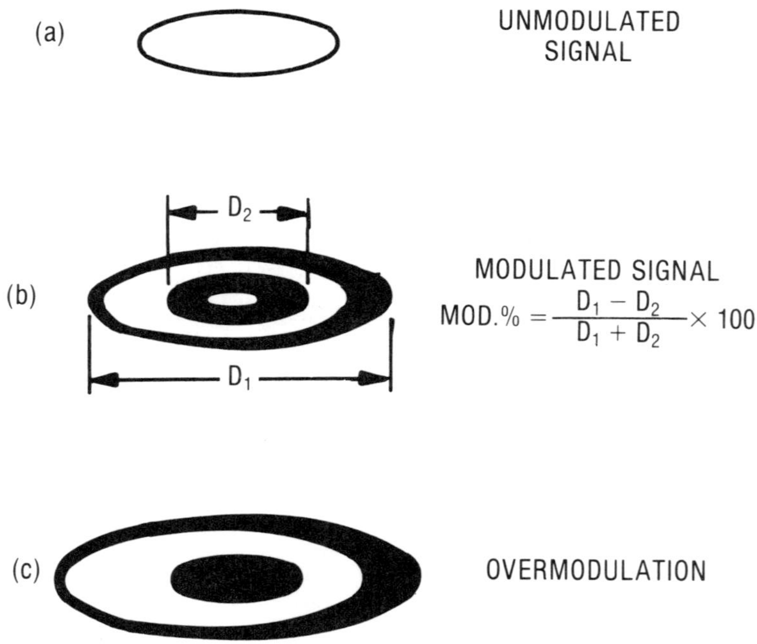

Figure 6.10 Typical Lissajous patterns

If the carrier is unmodulated, the scope will display an ellipse as in Figure 6.10(a). Modulation broadens the ellipse into a pattern like that in (b). The modulation index is calculated by measuring D1 and D2;

$$\frac{D_1 - D_2}{D_1 + D_2} \times 100 = \text{modulation percentage as before.}$$

In the third method, RF signals may be displayed directly on the scope by applying the signal to the vertical input and selecting a suitable sweep speed. This should be equal to, or a multiple of, the modulation frequency. The display, as shown in Figure 6.11(a), is a replica of the RF wave. The scope connections are shown in Figure 6.11(b).

We'll return to this subject again in the chapters on alignment and waveform analysis.

Measuring D.c. Levels

Since most vertical amplifiers are d.c. coupled, they may be used for measuring d.c. The input switch must be set to d.c., of course, and a free-running trace, of any sweep speed, is required on the screen. Generally, the "auto" position of the triggering mode control will provide this.

If you know the polarity of the voltage, you can position the trace on the bottom line of the graticule for positive voltages, or on the top line for negative voltages. Place the AC-GND-DC switch to GND when doing this to ensure a true zero, then return the switch to DC.

Also, if you have any idea of the value of the voltage, you can set the VOLTS/DIV control to a position that will allow for this value. For instance, if you anticipate that it will be around ten volts, you would select 2 V/div because that would give a deflection of around five divisions, the maximum deflection without going off the screen.

When you place the probe tip in contact with the voltage source the trace will jump from the zero position to the level corresponding to its value. This is measured in divisions of the graticule, and multiplied by the VOLTS/DIV setting.

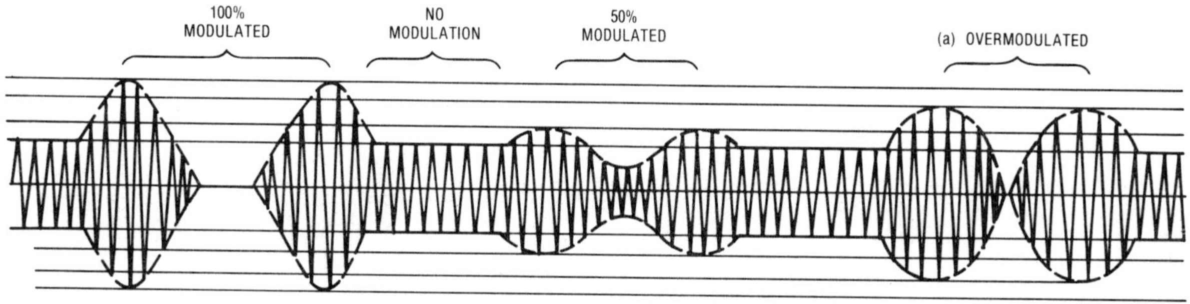

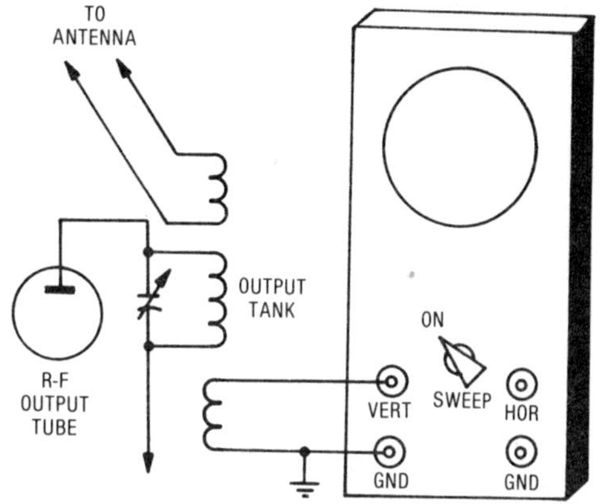

Figure 6.11 (a) Typical wave envelope patterns. (b) Connections for wave envelope display

Signal Generators and Their Signals

Before going further we should take time to review the signal sources available for obtaining the test signals for circuit analysis using waveforms. Test signals include *continuous waves and pulses*. *Continuous waves* are those that continue their periodic excursions without interruption. Typical continuous waves shown in Figure 6.12 (page 140) are:

Sine waves	Sawtooth waves
Triangle waves	Square waves

Pulses are individual excursions from a baseline with intervals between them called *baseline dwell times*. Typical pulses shown in Figure 6.13 (page 141) are:

Trapezoidal	Spiked
Rectangular	Rounded

Generators

The sources for these signals are:

Oscillators and Function Generators	Pulse Generators
Signal and Sweep Generators	

The *oscillator* generates sine-wave signals of known frequency and amplitude. If it is called an *audio oscillator,* it has a frequency range from 10 hertz to 20 kilohertz or higher. An *audio-signal generator* has a similar range, but higher power. A *test oscillator* is also an audio oscillator with a greater frequency range. The last two usually have meters to show the amplitude of the output signal.

Function generators are oscillators that produce triangle, sawtooth or square waves in addition to sine waves. (There are also generators that produce only one each of these functions.) They operate at frequencies from 100 kilohertz down to less than one hertz.

Signal generators are oscillators that produce higher frequency signals that can be modulated. These include RF signal generators (50 kilohertz through 40 gigahertz) with AM, FM or pulse modulation capability.

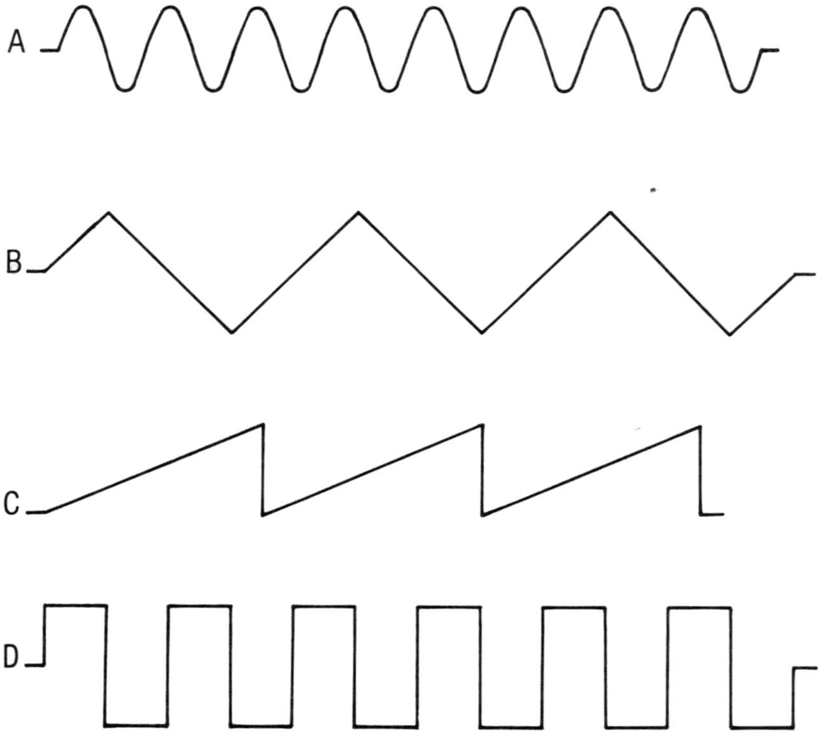

Figure 6.12 Continuous waves (a) Sine wave; (b) Triangle wave; (c) Sawtooth wave; (d) Square wave.

Sweep generators are an extreme case of frequency modulation. They put out a signal that continuously increases its frequency at a steady rate, starting at a lower limit and sweeping up to an upper limit, from which it returns at the same rate to the lower limit to start the next sweep. These are usually adjustable. Most have frequency markers. We shall discuss sweep generators again in the chapter on TV and radio alignment.

Pulse generators produce high-quality rectangular test pulses, with repetition rates mostly not exceeding one megahertz. The most important characteristics in a pulse generator are the *rise time* and *shape* of the pulse.

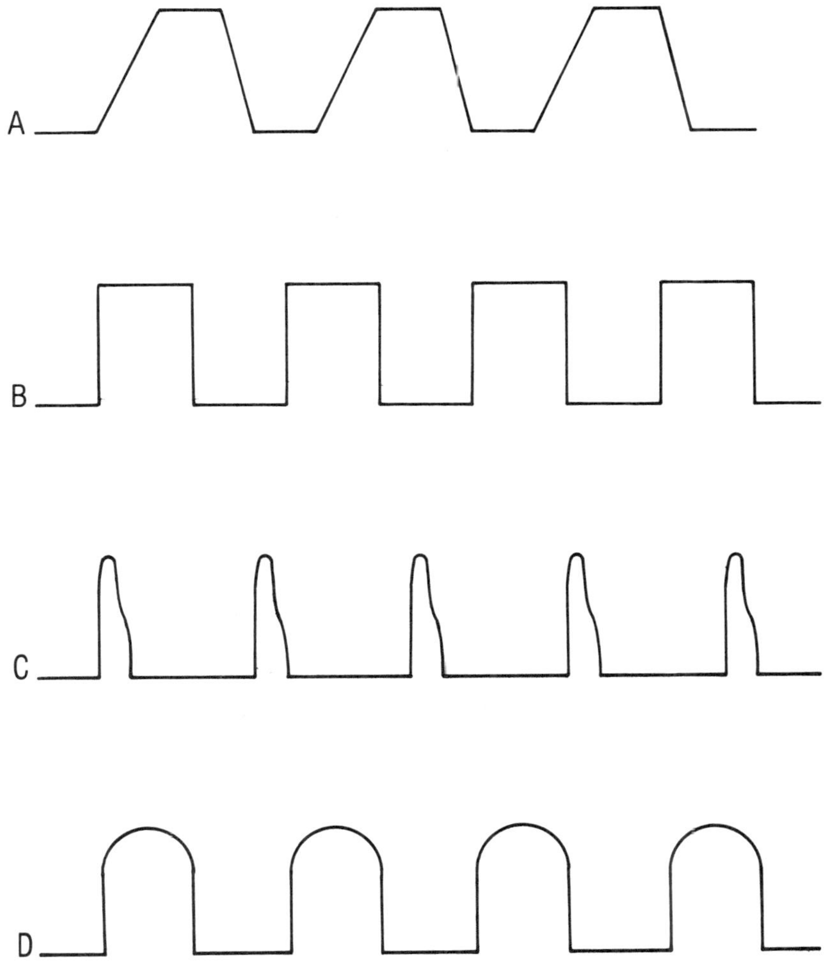

Figure 6.13 Pulses (a) Trapezoidal pulses;
(b) Rectangular pulses; (c) Spiked pulses; (d) Rounded pulses

Important Features in Oscillators

The most important features of oscillators are their:

Frequency range | Dial resolution

Output voltage or power	Harmonic distortion
Stability	Hum and noise

The *frequency range* of an oscillator is the first feature that determines whether it will be of any use to you for the purpose you have in mind. Obviously, you want one that will cover the frequency band you're interested in, with a reasonable margin on either side.

The *output voltage or power* of an oscillator depends upon the type of oscillator and the load placed upon it. Most oscillators are designed to give a certain output with a particular output impedance, such as 600 or 50 ohms. There also are some in which the impedance may be varied.

Stability is important for both frequency and amplitude. Frequency stability is obtained by using large amounts of negative feedback, precision components and carefully regulated voltages. Since temperature also affects frequency, an oscillator must be allowed an adequate time to warm up to its operating temperature when first turned on, after which it should not change appreciably. The amplitude should not alter as the frequency is varied, but remain constant over the entire range of the oscillator.

Dial resolution is a measure of how close the oscillator's frequency at its output terminals is to the frequency indicated on its tuning dial.

Harmonic distortion occurs because harmonics of the fundamental frequency are also generated, and can degrade the purity of its waveform. Some harmonic distortion is always present, but as long as it is only some fraction of one percent it is negligible.

Hum and noise must be kept as low as possible for the same reason. This requires careful shielding, grounding and filtering.

All signal sources are operated in much the same way. The following step-by-step procedure applies to low-frequency generators. Some additional procedure is required for higher frequencies.

(1) Unless battery operated, connect to suitable power source and turn power switch on. Allow sufficient time for oscillator to stabilize. This should be about 30 minutes for vacuum-tube equipment, less for other.

(2) Set range and frequency controls for desired frequency. Frequency dial setting must be multiplied by range switch setting. For instance, 45 (on frequency dial) × 10 (range setting) = 450 hertz.

(3) Set function switch (if applicable) for the waveform you want.

(4) Connect equipment under test to output terminals of oscillator, using matching pad or attenuator as necessary. (If a very small output voltage is required, it is better to use an external attenuator to reduce the ratio of hum and noise.)

(5) Adjust amplitude control for output voltage required. This may be monitored at the input of the unit under test with an AC voltmeter if accuracy is required.

(6) The output terminals of low-frequency oscillators and generators consist of two or three binding posts. One of them (often colored red) is for the output from the high side of the output amplifier, the other is the ground return and connects to the chassis. If the output amplifier is a differential amplifier (see Chapter 3), there will be an additional terminal, because the amplifier has two high sides. This terminal may be connected to ground with a shorting strap between it and the ground terminal, to give an unbalanced output, or not so connected, to give a balanced output.

    These terminals are normally spaced ¾-inch apart so you can plug in a "double-banana" plug.

For higher frequencies, and especially to preserve the shape of fast-rising pulses and square waves, you have to take additional steps to make proper connections between your generator, the circuit under test and other equipment connected to them. The higher the frequency and the faster the risetime, or the longer the connecting leads, the more important this is. Shielded or coaxial test leads, with suitable matching pads, as described in Chapter 8, are definitely required.

7

Using Your Scope for Troubleshooting

Signal tracing is a *dynamic* testing procedure. Dynamic tests are tests performed under actual operating conditions, as opposed to *static* tests, which are made with the equipment idling or turned off.

Dynamic tests may be used to isolate all types of troubles, including not only serious defects causing major changes in equipment operation, but also minor problems affecting circuit efficiency and performance, but which allow nearly normal operation.

Signal tracing is the most powerful dynamic testing technique, and the scope is the most versatile of the test instruments you can use. Signal tracing involves tracking a signal stage by stage as it passes through the equipment. In Figure 7.1 there is a composite block diagram that can be applied to virtually any type of electronic equipment. As you can see, a signal source applies a suitable signal to the input device. The indicating device—your scope—then checks the relative amplitude and quality of the signal at the input and output of each stage. These are the points marked A, B, C, D, E, and F in Figure 7.1. Each stage, when operating normally, is designed to do something to the signal. If you find that it is doing so correctly, then that stage is working properly. If your scope shows it failing to do so, then you have found the defective stage.

The designations given to the blocks in Figure 7.1 are general ones. In the actual equipment they will vary. Some examples are listed in the following table:

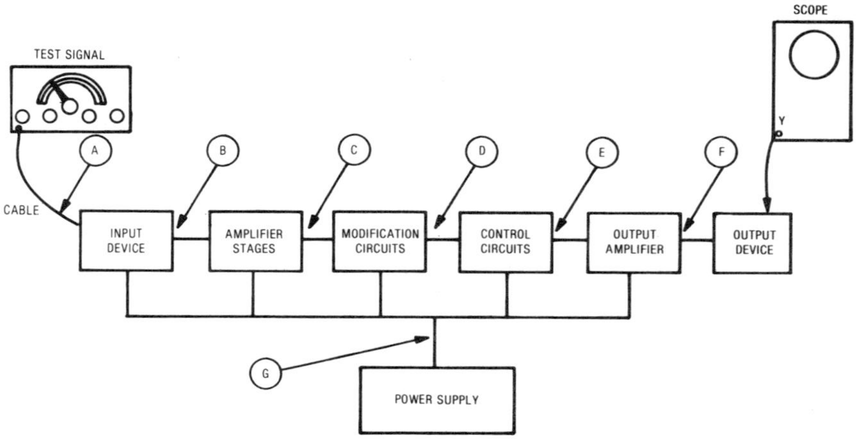

Figure 7.1 This composite block diagram can be applied to almost any type of electronic equipment.

Signal Tracing in TV, Radio, Audio and Logic Circuits

Using the signal tracing technique may be summarized as follows:

(1) Turn the equipment on.

(2) Apply an adequate test signal to the equipment's input. In a receiver, you would obtain this signal from an RF signal generator. In an audio amplifier, this signal may be obtained from a record player, tuner or audio generator. In other cases, depending on the equipment tested, the signal may be obtained from a pulse generator, square-wave generator, sawtooth oscillator, or other device. If the equipment generates its own signal, as in a radio transmitter, a separate source may not be needed.

(3) Make sure the connection of the signal source does not disturb the circuit. Where necessary, insert a d.c. blocking capacitor in series with the input lead.

(4) Make sure that the test signal does not overload the equipment. Use the minimum signal needed for usable output.

TYPE OF EQUIPMENT	INPUT DEVICE	AMPLIFIER STAGES	MODIFICATION CIRCUITS	CONTROL CIRCUITS	OUTPUT AMPLIFIER	OUTPUT DEVICE
Public address	Microphone	Preamp.	Tone controls	Gain control	Power amp.	Speaker
Radio receiver	Antenna, R.F. Front End	I.F. amp.	Second detector	Volume control	Audio section	Speaker
Radio transmitter	Crystal oscillator	Buffer amp.	Modulator	Impedance matching network	Final amp.	Antenna,
TV Receiver (video only)	Antenna Tuner	Video I.F. amp.	Video detector	Brightness, contrast controls	Video amp.	Picture tube

(5) With your scope, check the relative amplitude and quality of the signal at the input and output of each stage.

(6) The test signal should be modified by each stage. In the case of an amplifier stage, for instance, the signal should be increased in amplitude. In a clipper stage a portion of the signal may be removed, or "clipped" off.

(7) If the signal is changed in an unexpected fashion; for example, if the amplitude drops when it passes through an amplifier stage, trouble is indicated, and the defective stage has been isolated.

(8) Where necessary, change the type of probe as the signal is followed through the equipment. For instance, use a demodulator probe in checking the RF and IF stages of a receiver, and a passive probe in the audio section.

In addition to isolating a defective stage, the signal tracing technique may be used for other jobs. An open coupling capacitor may be identified by comparing signal amplitudes on each side of the unit. An open bypass may be found by checking across it for the presence of a signal. Stage gain may be measured by comparing the relative amplitudes of the input and output signals, taking circuit impedances into account. You can also track down undesired signals such as hum, noise, interference, oscillation, and so on.

In detecting minor changes in equipment performance, waveform analysis is used. We have already touched on this, and will do so again, especially in Chapter 9. In this technique a signal of accurately known characteristics is used, and its waveform is examined at the input and output of each stage. Some types of equipment, especially television receivers, have waveforms that are characteristic of certain circuits or components, and these are shown in their schematic diagrams. Comparison of these with the scope display may pinpoint trouble if a significant difference is seen.

Signal Injection

Signal injection is also a dynamic test technique. It is complementary to signal tracing, and is almost as powerful for tracking down trouble; in some equipment it is even more effective. As the name indicates, this technique involves "injecting" a test signal into the equipment, starting with the input to the last stage, and working backward, stage-by-stage, to the input to the first stage. The scope is connected to the output of the last stage all the time.

Signal Injection in the Video Amplifier

You can test the performance of a video amplifier with an injected signal from a square-wave generator. Connect your scope and generator to the amplifier as shown in Figure 7.2. Set the scope controls as follows:

SEC/DIV	.5m
TRIGGERING SOURCE	INT
TRIGGERING MODE	AUTO
VOLTS/DIV(X10 PROBE)	1
AC-GND-DC	AC

On the TV receiver, disconnect the antenna and turn the channel selector to an unused channel. Remove the video detector, if it is a tube. Otherwise, disconnect the signal input to the video amplifier.

Connect the square-wave generator output to a point before the input coupling capacitor. If there isn't one, make the connection through a 0.1 μF capacitor.

Connect the scope's compensated probe to the output of the square-wave generator.

Turn all the equipment on, and give it sufficient time to warm up and stabilize.

Adjust the square-wave generator for an output of 400 Hz with an amplitude of 5 V peak-to-peak, as seen on the scope.

Adjust the compensated probe, if necessary, for a good square-wave with flat top and right-angled corners.

Transfer the probe to the output of the video amplifier.

Adjust the contrast control and the scope VOLTS/DIV control for a display of convenient size, but do not overdrive the amplifier.

The waveform displayed on your scope should be very close to the square-wave observed at the input. The 400-hertz square-wave gives you an indication of amplifier frequency response from 40 to 40,000 hertz. If there is a decided slope at top, bottom, or both, there is something amiss with the coupling capacitor or input resis-

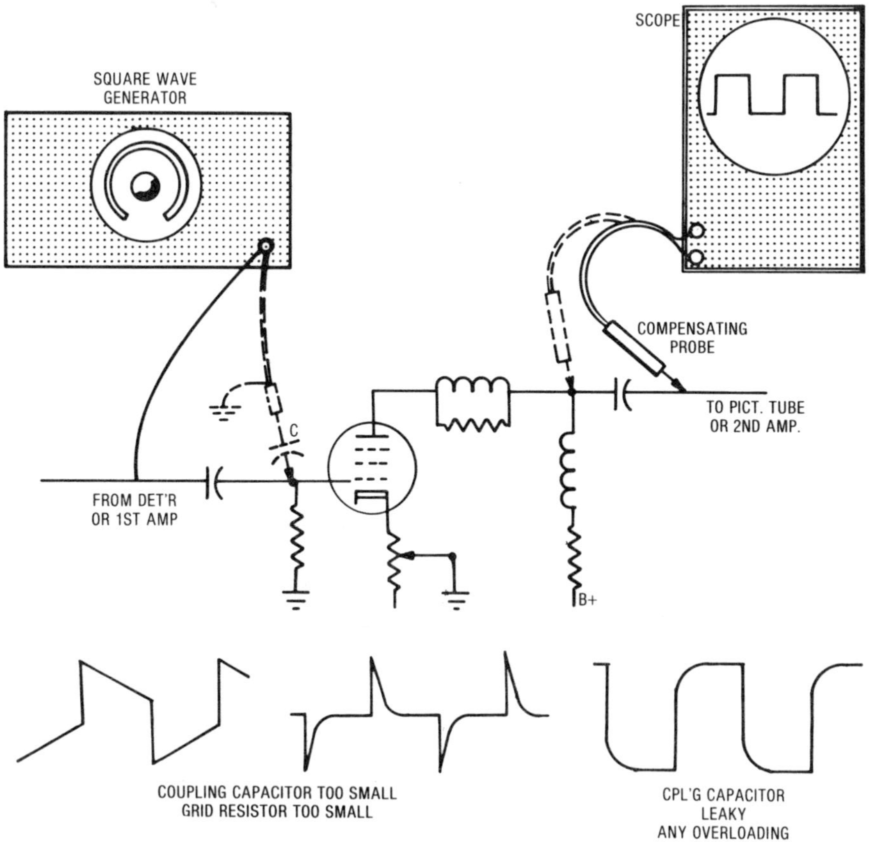

Figure 7.2 Square wave test setup for a video amplifier and some traces which indicate particular faults

tor (either is too small). Very sharp peaks indicate very severe faults in this area. Rounded corners, on the other hand, may be caused by a leaky coupling capacitor. (They can also be caused by overdriving the video amplifier.)

If there are two stages in the amplifier, and the overall response is unsatisfactory, transfer the scope probe to the output of the first stage to see if the indications originate in that stage. If they do not, you know the cause of the trouble must be in the second stage.

Frequency response may also be examined by using a sweep generator as explained in Chapter 8.

Signal Tracing in the Sync Section

In this procedure we use the actual sync signal derived from the composite video signal broadcast by the TV station. Any channel will do. The only requirement is that the station signal be of sufficient strength to give a good picture on a set that is operating properly.

Temporarily disable the vertical sweep oscillator and horizontal a.f.c. stage to prevent possible problems in these sections from affecting the sync section. (In a tube set this may be done by removing the appropriate tubes. In others you will have to disconnect the input leads.)

Connect your scope probe to the output of the sync section. The waveform should have the general appearance of Figure 7.3, but check the manufacturer's schematic for the test point to be used and characteristic waveform for the model you are troubleshooting. If it has the proper shape, your problem is not in the sync section.

However, if the waveform is unsatisfactory, transfer your scope probe to the input to the sync section and look at the composite video signal. (You may have to disable the sync section temporarily to be sure any defect in the video signal is not being caused by the sync section.) However, if the video signal has the proper appearance, then your problem must be in the sync section.

Many scopes designed for TV servicing have an additional triggering mode position labeled "TV." This setting synchronizes the sweep with the vertical or horizontal frequency, so you should use it if you have it.

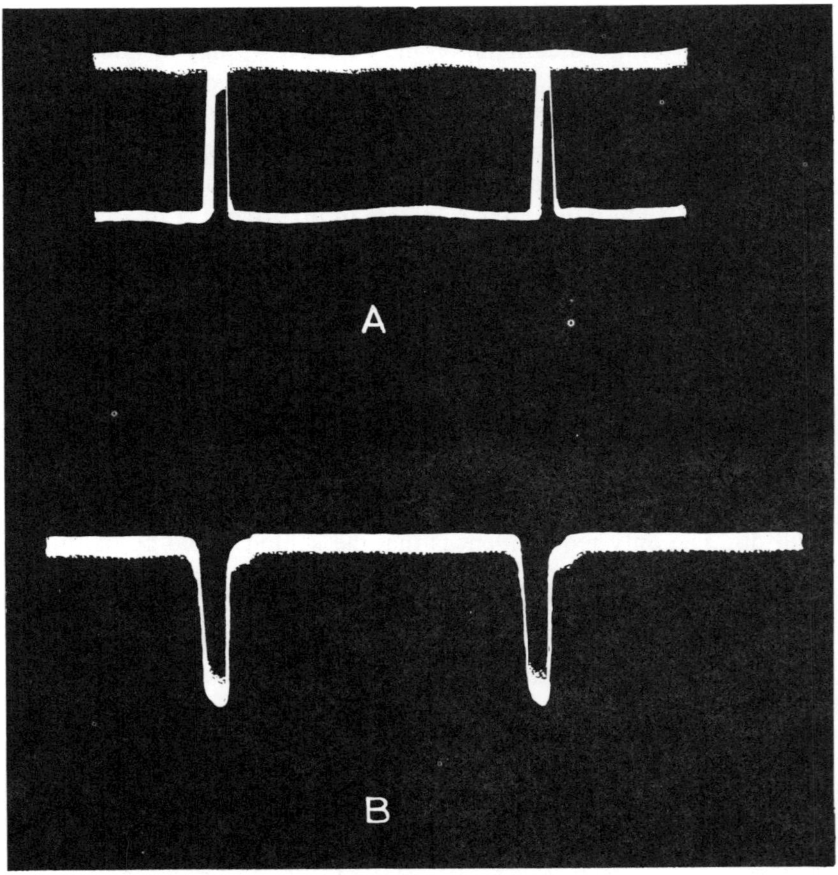

Figure 7.3 Output waveforms of Sync Section (a) Horizontal pulses; (b) Vertical pulses.

Signal Tracing in the Vertical Sweep Section

For a quick check of the vertical sweep section connect your scope probe to the following points:

1. If the circuit employs a plate-coupled multivibrator or a blocking oscillator, connect to the input after the integrating filter.

2. If the circuit employs a cathode-coupled multivibrator, connect your scope to the grid or base of the second section.

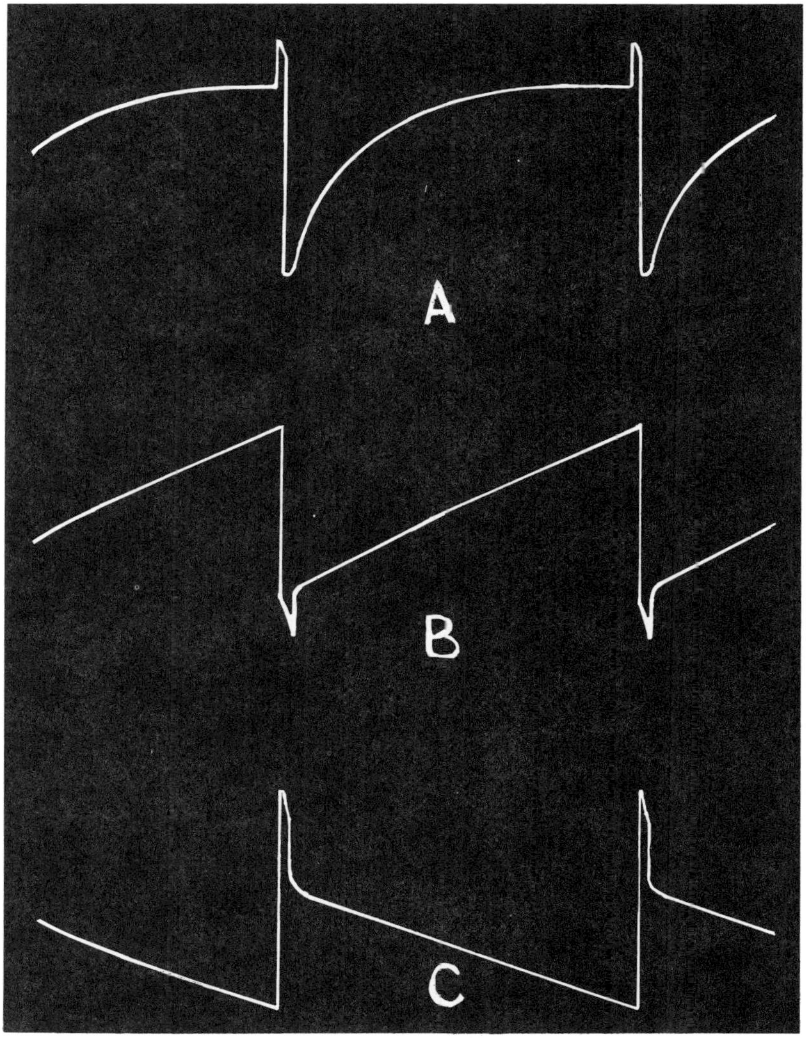

Figure 7.4 Vertical Sweep Waveforms
(a) Input signal to oscillator
(b) Output sawtooth from oscillator
(c) Output signal of amplifier

The input signal should be as shown in the manufacturer's schematic, or of the general form of Figure 7.4(a).

Now connect your scope probe to the output of the multivibrator or blocking oscillator.

1. With a plate-coupled multivibrator, this will be the output of the first section.
2. With a cathode-coupled multivibrator, this will be at the output of the second section.

The output sawtooth signal should be as shown in the manufacturer's schematic, or of the general form of Figure 7.4(b). This waveform should also be found at the grid or base of the following tube or transistor. The waveform at output of this stage goes to the vertical deflection coils, and should resemble Figure 7.4(c).

When following through the vertical sweep section in this manner, waveforms will be normal until the defect has been passed, so look for the trouble between the point at which the waveforms first become abnormal and the last point at which they are normal. The waveforms do not have to be exactly like the illustrations, but should be generally similar and free from noticeable peculiarities or unusual fuzziness (except at the yoke coils).

The pulse voltages at the plates of vertical output tubes may be

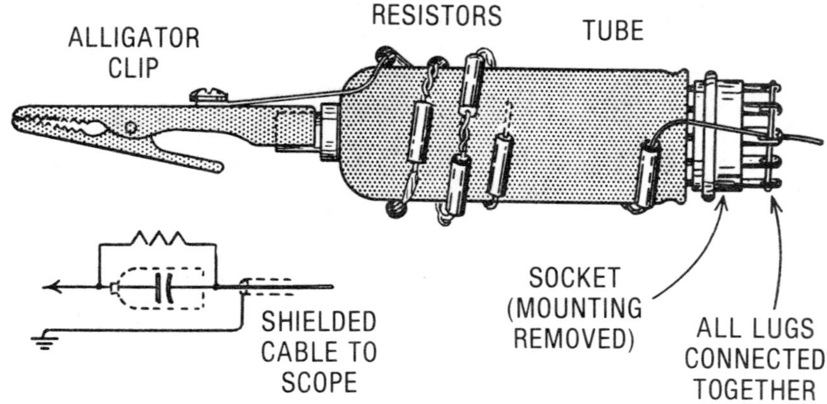

Figure 7.5 Construction of a probe for a capacitance voltage divider used for observing high-voltage waveforms. Tube should be a 1X2A or B. Total resistance should be 40MΩ approximately.

too strong to be applied safely to the scope input. If this is so, you will need a capacitance voltage divider probe, with a step-down ratio of 100 to 1. Alternatively, you can make one as shown in Figure 7.5.

Signal Tracing in the Horizontal Deflection Section

The horizontal deflection section consists of the circuits shown in Figure 7.6. There are two inputs to this section: from the sync section, and feedback from the sawtooth oscillator. These are applied to the automatic frequency control (AFC) circuit, which compares the two signals and applies the correcting voltages to maintain the oscillator frequency at the correct rate. It operates like a governor on a steam engine, varying the "throttle opening" as required to keep the "r.p.m." constant. Consequently, you should begin by investigating these inputs when you have troubles in this section.

If waveforms from the sync section are satisfactory you can eliminate it as the cause of the trouble. If sawtooth waveforms from

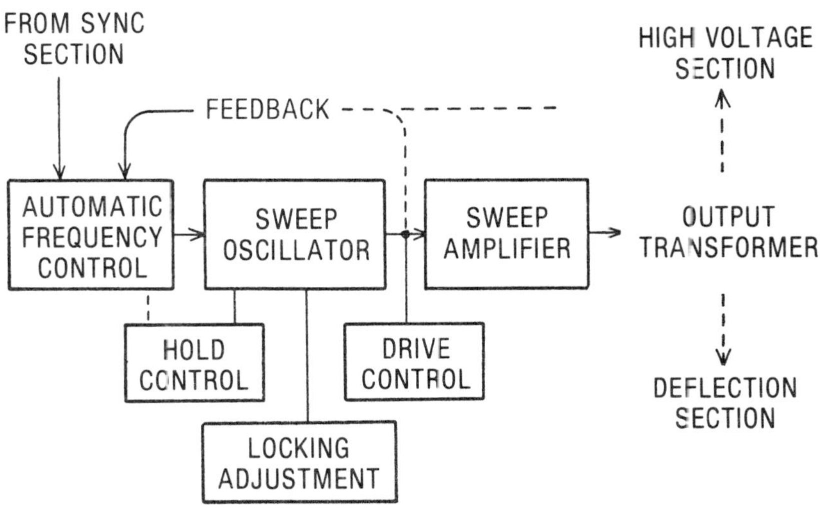

Figure 7.6 The horizontal sweep section extends from the sync output to the horizontal output transformer

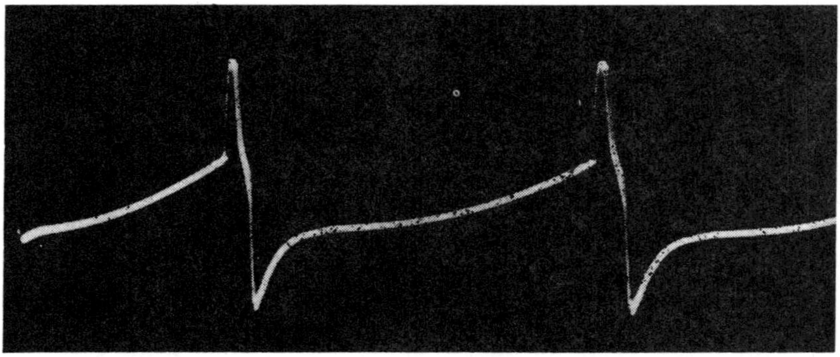

Figure 7.7 Typical waveform when both horizontal AFC signals go to the same input.

the oscillator are arriving at the AFC input, you can assume the oscillator is working. But both must be present, of course. If there is only a sawtooth or approximate sawtooth, sync pulses are not reaching the AFC section (or are not of the correct waveform). If there are only separated sync pulses, without any sawtooth, the oscillator is in trouble. The proper input should be both signals, which in some AFC circuits go to separate inputs, but in others share the same input to give a combination signal, as in Figure 7.7.

The following table shows the three main types of oscillator used in horizontal deflection circuits, and the type of AFC most likely to be used with each one.

OSCILLATOR	AFC
Multivibrator	Phase detector
Blocking	Pulse width
Hartley	Reactance tube (controlled by phase detector)

Separate inputs for sync and sawtooth signals are provided for phase detectors, but pulse width AFC circuits use the same input for both. The sawtooth input to the AFC circuit is, of course, the same as the output waveform of the oscillator that is applied to the input of the horizontal sweep amplifier.

Waveforms in the output portion of the horizontal deflection section may vary widely in different receivers, so it is best to compare them with the manufacturer's data. In view of the higher voltages prevailing, be sure to use a high-voltage probe such as the capacitance probe described previously.

Ringing Test of Flyback Transformer

A useful test of the condition of a flyback transformer can be made by connecting your scope and a square-wave or pulse generator to it, as shown in Figure 7.8. The "gimmick" is a small capacitor or a few turns of wire wound around one of the transformer leads. Connect a capacitor in parallel with the transformer, if there is normally one in the position in the circuit. The value should be about the same. Each pulse will give a display called a ringing waveform, if the transformer is good. It is possible to estimate the Q of the transformer by adjusting the amplitude of the first cycle of the waveform to fit between the ten vertical divisions of the graticule as shown. Then note the point where the waveform has decreased to 1.85 divisions above and below the centerline (this is the 37-percent level). Count the number of peaks from the beginning of the waveform to this level (that is, the number of peaks from the 100-percent level to the 37-percent level), and multiply this number by π (3.1416) to obtain the value for Q. Around 30 is a typical value. This test can be used for various radio and television inductors, as well as for flyback transformers.

Some service scopes provide a pulse output for performing this test. In other cases you can use a calibrator output, if it is a square wave. The time base should be adjusted to display one pulse or square wave on the screen so you get a single ringing waveform.

Signal Tracing in the Color Section

With the color-bar generator and scope connected as in Figure 7.9, and the TV receiver's tint or hue control set to midpoint, you can observe the output of each demodulator. The R-Y demodulator should give a display as in Figure 7.10(a). The important thing to

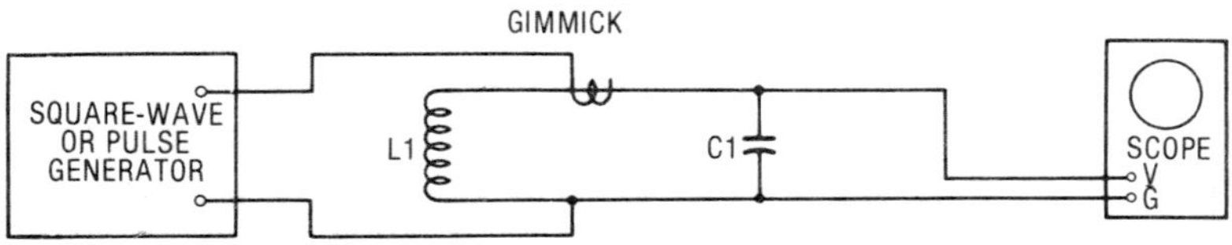

(A) TYPICAL TEST ARRANGEMENT.

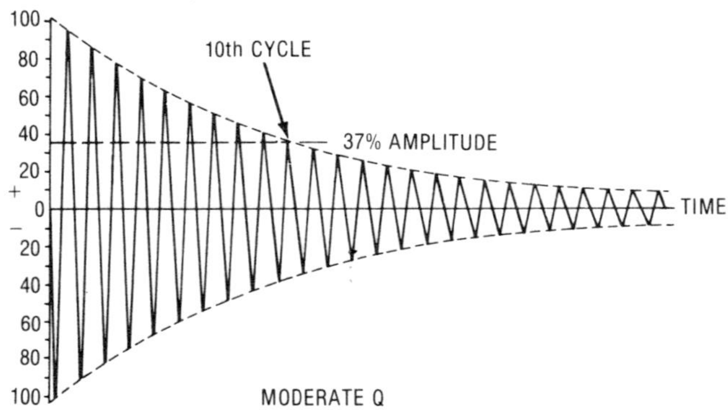

(B) ANALYSIS OF A WAVEFORM TO MEASURE THE Q VALUE.

Figure 7.8 How to perform a ringing test of an inductor (L1), such as a flyback transformer. The scope sweep rate should be adjusted to display one waveform, and the amplitude of the left-hand edge of the display should occupy five divisions above and below the horizontal center line of the graticule. C1 should have a value roughly equal to the capacitance paralleling L1 in the circuit that it goes in.

note is that the "curve" crosses the zero line between color bars 5 and 6 (reddish-blue and blue).

When you apply the scope probe to the B-Y demodulator output you should get the display in (b). This is phase-shifted 90 degrees from that in (a), and crosses the zero line at two points: between color bars 2 and 3 (orange and red) and 8 and 9 (cyan and bluish-green).

If the signals in Figure 7.10 are incorrect (and the color-bar pattern on the TV screen is also incorrect), you must adjust the color sync tuned circuit going to the R-Y or B-Y demodulator to obtain the proper display on the scope. This may not at this stage produce the proper color-bar pattern on the picture-tube screen. Other adjustments may be required also. But as long as the color-bar

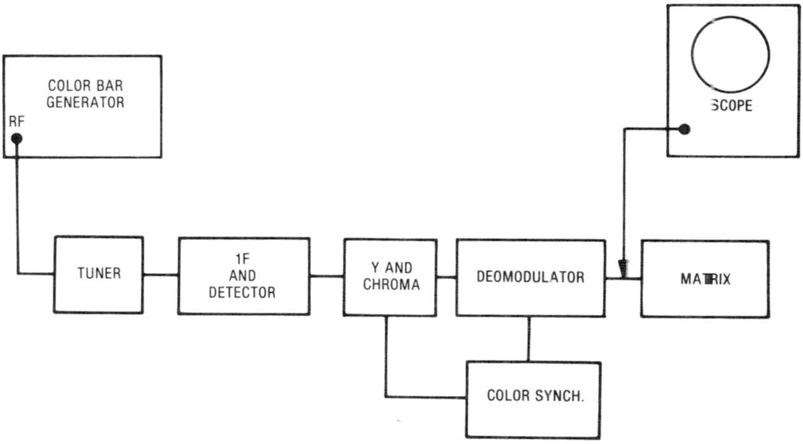

Figure 7.9 How to connect the color-bar generator and scope for signal tracing in the color section.

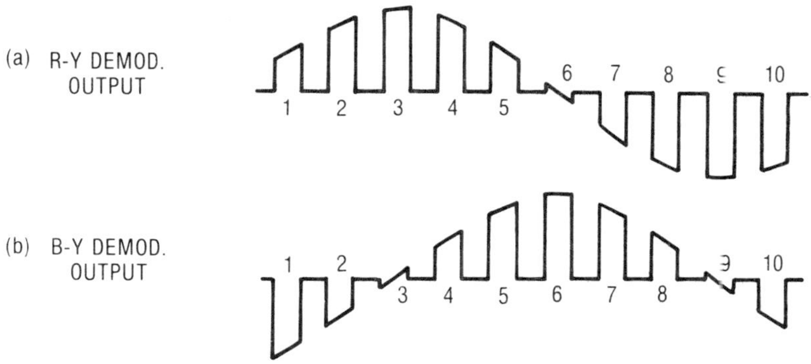

Figure 7.10 R-Y and B-Y demodulator output waveforms

generator has the standard sequence of color bars and the demodulator outputs appear as in Figure 7.10 you have got the R-Y and B-Y channels aligned correctly.

The proper sequence of color bars as seen on the picture-tube screen, reading from left to right, is:

1. Yellow-orange
2. Orange
3. Red
4. Magenta
5. Reddish-Blue
6. Blue
7. Blue
8. Cyan
9. Bluish-green
10. Green

If you want to check the amplitudes of the three color-difference signals supplied by the matrix circuits, you use the same setup as in Figure 7.9, but connect the scope probe to the R-Y control grid of the picture tube, and synchronize the scope sweep with the color-bar generator.

Set the chroma gain control to its midway position and adjust the hue or tint control until the red color bar (3) is at its maximum amplitude on the scope screen. Mark this point on the scope screen, or note the number of vertical divisions.

Reconnect your scope probe to the B-Y control grid of the picture tube, and repeat this procedure for the blue color bars (6,7). If the matrix is operating correctly, the maximum amplitude of the blue color bars should be about the same as the red.

Reconnect the scope probe to the G-Y control grid, and repeat the procedure for the green color bar (10).

These color bars should have similar amplitudes, indicating approximately the same amount of gain through the matrix. If one is substantially larger or smaller, it may be due to a defect in that particular matrix section. However, check the output of the corresponding demodulator to make sure it is the matrix.

Ordinarily, the matrix circuits of modern color receivers do not have any adjustments. Defects are also comparatively rare. In the unlikely case of such a defect you will need to locate the problem component by voltmeter and ohmmeter tests.

8

Alignment of TV and FM

Alignment of TV or FM circuits is required when they do not give the proper amplification to all frequencies within their pass band. The performance of the circuit is demonstrated by displaying its *response curve* on a scope. This curve is produced by applying a band of frequencies to the scope from a *sweep generator* and observing the result.

Sweep Generator

The sweep generator supplies a band of frequencies 10 to 15 megahertz wide. For example, you could apply the range of frequencies between 40 and 50 megahertz to the input of a TV i.f. amplifier. This would result in a scope display of the i.f. amplifier frequency response over that band, and comparing this with what it should look like will tell you at once whether or not alignment is necessary.

The sweep generator does not supply this entire range of frequencies simultaneously. It first produces a 40-megahertz signal, and then gradually increases the frequency to the upper frequency limit of 50-megahertz. Then the cycle reverses, and the signal frequency decreases to 40 megahertz. This increase of frequency from 40 to 50 megahertz and decrease to 40 megahertz again is repeated 60 or 120 times a second.

Sweep generators have two outputs. The RF output is connected to the input of the unit being tested. The sweeping voltage output is connected to the horizontal or X input of the scope. This supplies a triangular waveform to the horizontal amplifier to drive the electron beam back and forth across the screen in step with the increase and decrease of the signal frequency. The vertical input of the scope is connected to the output of the unit being tested.

In order to judge accurately the frequency of any point on the response curve you also need a *marker generator*. This is built into some sweep generators; with others you need a separate instrument. The marker generator produces a small "pip" on the response curve. The marker generator is designed to generate pips with precisely known frequencies, so you can tell the exact frequency of the point on the curve where the pip occurs. You can set the controls of the marker generator to give a pip at the frequency of, say, the picture carrier, the color subcarrier or the sound carrier, to see if they are positioned correctly, or if the curve has the proper form at those points.

You can add the marker to the response curve in various ways. If the sweep generator has an internal marker system, the marker signal can be inserted into the sweep signal by throwing the proper switch and adjusting the marker amplitude control. If the marker generator is a separate instrument, you should apply it in accordance with the instructions of the manufacturer of the unit you are aligning. Alternatively, connect it at the same point as the sweep signal or some other point where there is less loading or interaction.

This is most important. You must connect the marker generator at some point where it will not load or distort the response curve. Further, you must keep the amplitude as low as possible, until you can just see the pip, to avoid changing the shape of the curve. Usually you should apply the marker through a blocking capacitor.

Test Equipment Setup

There are, of course, various ways of connecting your test equipment to the unit to be aligned. The most common arrangement is shown in Figure 8.1. Most manufacturers give detailed instructions on how to align their equipment, and it is always best to follow these rather than proceed in a hit-or-miss manner. But whatever procedure you follow, the four provisions listed below are very important:

a. A proper ground system.
b. Accurate bias and signal levels.
c. Adequate shielding.
d. Proper injection of marker signals.

The type of ground system often makes the difference between a satisfactory response curve and a variable and indefinite one. With a good ground system, your test equipment will show a true response that will not vary with placement of interconnecting leads and changes in test equipment position.

You must follow the manufacturer's recommendations step by step to get the response curves shown as typical in his instructions. The amplifier must be operated with the specified bias voltages, and the signal and output levels must be of the required values. If these are incorrect a distorted response pattern will be the result. With minor variations, all manufacturer's alignment procedures include the following:

a. Adjust contrast and a.g.c. for proper bias and levels.
b. Apply an external or a controlled internal voltage to the bias line to set up correct operating conditions.
c. Apply a signal with as low an amplitude as will give you a reasonably sized display. If the amplitude of the signal is too high it may overload the receiver, causing the response curve to be flattened.
d. Take precautions when connecting test equipment to the unit being aligned so that the equipment does not influence circuit operation enough to upset the response curve.

It is always important to use a logical step-by-step procedure in adjusting the circuits in alignment procedure. First, the test equipment must be interconnected correctly. Second, all the recommended precautions must be taken to obtain reliable response indications. Last, do not make random adjustments of the controls, and avoid using hit-or-miss tactics. Follow the recommended procedures exactly, and don't make any adjustment until you are sure you know what the effect of the adjustment will be.

And finally, perform alignment only as a last resort, because

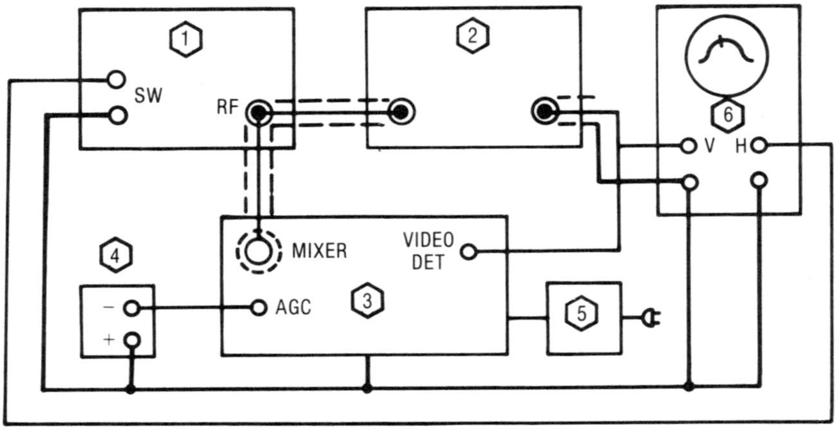

Figure 8.1 How to connect your test equipment for TV alignment.

1 *Sweep Generator:* using 75-ohm coaxial leads, connect RF output to ungrounded tube shield over mixer tube in tuner, or to test point provided in transistorized tuner; and also to marker generator input. Connect test leads from sweep output (60 Hz) to oscilloscope horizontal input (H). Adjust sweep generator frequency to 44 MHz (10 MHz sweep).

2 *Marker Generator:* connect output to oscilloscope vertical input (V). This is called "post injection," and avoids having the markers go through the circuits being aligned.

3 *Television Receiver:* connect bias box to AGC bus, and adjust for bias voltage specified. Connect lead from video detector output to vertical input (V) of oscilloscope. Connect to power line using isolation transformer. Allow 20 minutes warm-up.

4 *Bias Box:* nine-volt battery with potentiometer.

5 *Isolation Transformer:* 1 : 1 power transformer for safety.

6 *Oscilloscope:* Adjust controls for stable display similar to that shown. If signal is noisy conncct a 47-ohm resistor in series with vertical input and lead from video detector, with .05-microfarad capacitor from vertical input to ground.

---

modern circuits seldom need it. If the response curve is distorted, the trouble is far more likely to be caused by a defective component than by misalignment, so look for defects first.

Sometimes you may want to ''touch up'' the alignment of a receiver rather than carry out the complete alignment procedure. When doing this, make only one adjustment at a time. Before moving the control, note where it is set. Then move it on either side of this setting to see if there is any improvement in the response.

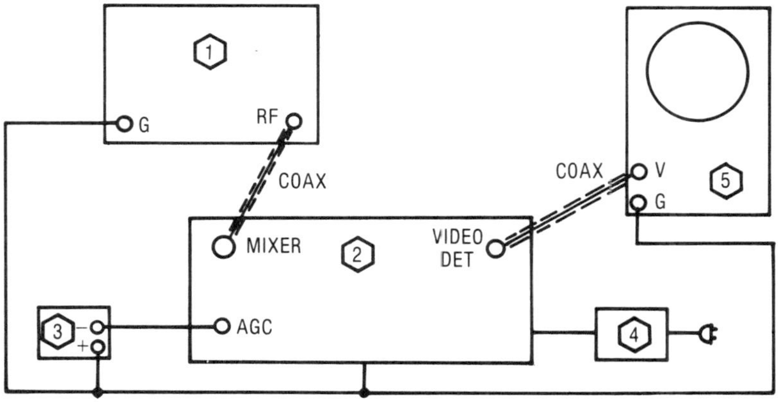

Figure 8.2 How to connect your test equipment for i.f. trap alignment.

1 AM signal generator with output of 41.25 MHz modulated with 400 or 1000 Hz.

2 TV receiver disconnected from antenna and tuned to an unused channel.

3 Bias box adjusted for output of −3 V.

4 Isolation transformer.

5 Scope with sweep rate set to 1μs/div.

Before proceeding to another control, restore the first one to its original setting. This avoids the usual problem in making touch-up adjustments of getting several controls misadjusted, leading to a hopeless mess. When this happens, you'll have to go through a full alignment procedure to restore normal operation.

Procedure for Aligning an I.F. Amplifier

Connect the test equipment as in Figure 8.1. Connect the scope vertical input directly to the video detector load. It is not practical to connect a demodulator probe to the output of the last i.f. amplifier stage, because you would not get a true response unless all the characteristics of the probe were exactly like those of the video detector.

Trap Alignment

You should perform trap alignment before overall alignment. Use the manufacturer's procedure if available. (The following general procedure is given to supplement it, not substitute for it.)

1. Connect test equipment and unit to be aligned as in Figure 8.2. (Note that you are using an AM signal generator at this stage, as you need a single frequency, not a band of frequencies.)
2. Turn all the equipment on, and allow it time to stabilize before proceeding.
3. On the scope, set the TRIGGERING SOURCE to INT and TRIGGERING MODE to AUTO. Use a time base of 1 microsecond. (These are the controls illustrated in Figure 1.3 (Chapter 1). Allow for variations on different models.)
4. Adjust the bias box for an output of 3 volts.
5. Adjust the AM signal generator for an output of 41.25 megahertz modulated by 400 or 1000 hertz. Do not overload the circuit with too high a signal level, but adjust the scope VOLTS/DIV control for a display that uses most of the screen. (The display will be a broad band across the screen.)
6. Adjust each sound i.f. trap in turn (in some cases there may be one only). The amplitude of the scope display will decrease as you approach the minimum setting, increase as you go away from it. Adjust each trap for the minimum indication, increasing the scope sensitivity (VOLTS/DIV control) as necessary to maintain an adequate display. If the amplitude of the display gets too low to distinguish the minimum, reduce the output of the bias box. After each trap has been set for the minimum indication, go over them again for a final touch up.
7. Readjust the AM signal generator for a modulated output of 47.25 megahertz, and adjust the adjacent-channel sound trap(s), using the same procedure as for the sound i.f. traps.
8. If a trap for the adjacent-channel video signal is used, it is usually tuned to 38 or 39 megahertz. You adjust it in the same way as the other traps, but with the AM signal generator set to the required frequency.

Bandpass Alignment

Bandpass alignment is the overall alignment of the i.f. amplifier by adjusting the i.f. transformers for the proper response. The general procedure for this is as follows:

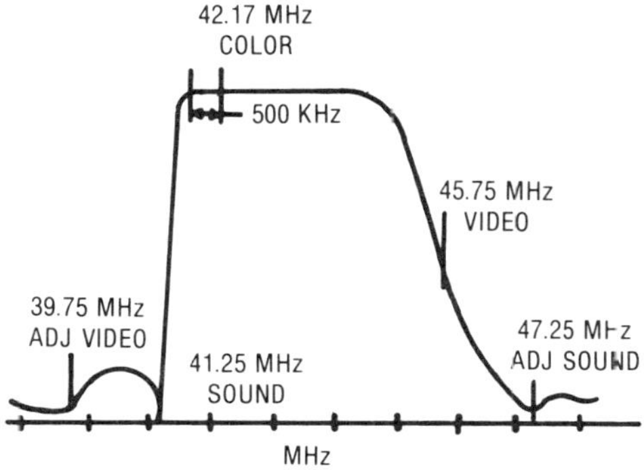

Figure 8.3 Ideal i.f. response curve

1. Connect the test equipment as shown in Figure 8.1.
2. Turn everything on and let it warm up and stabilize.
3. Set the sweep generator for maximum sweep width, and center frequency of 44 MHz.
4. Set the oscilloscope controls (using the scope in Figure 1.3 as a model) as follows:

X1-X10	X10
TRIGGERING MODE	X-Y
VOLTS/DIV	.1
AC-GND-DC	AC

5. Adjust the bias box for the voltage specified by the manufacturer.*
6. Set the sweep generator output amplitude to zero. Now increase it gradually until the response curve seems to overload. Reduce the amplitude slightly. The ideal response curve would appear as in Figure 8.3. Your aim is to obtain one from your i.f. amplifier that is a "reasonable facsimile" of it.
7. Set one marker to 45.75 MHz and the other marker to 42.2 MHz. If only one marker is available, mark the spot on the screen with a grease pencil where the 45.75 MHz pip appears, and reset the marker to 42.2 MHz.
8. The manufacturer's alignment procedure will show the fre-

*This will be the voltage that gives 5 or 6 V P-P video output with a signal input of about 2mV P-P.

quency to which each coil should be tuned. They are tuned to different frequencies ("stagger-tuned") to cover the i.f. bandwidth. If the response curve shows a hump where it should not, while a dip occurs elsewhere, you would assume that one coil at least was tuned to the wrong frequency. Touch up the various adjustments to get the response as near to that in Figure 8.3 as possible.

If some of the coils are seriously misaligned, you may have to do a stage-by-stage alignment before you can do the overall alignment. To do this, connect the sweep generator to the input of the last stage, and align the last i.f. coil for its correct frequency. Next, connect the sweep generator to the input of the previous stage and repeat for that circuit. When all coils have been adjusted, repeat the overall alignment given above.

In a black-and-white TV set, the procedure so far would be all that is necessary. In a color set, you have to make sure that no improper phase shift occurs, since this would affect the color reproduction.

To do this, vary the bias between 6 and 1 volts, adjusting the sweep generator and scope VOLTS/DIV control to keep the display to about the same size. The markers should stay at the same points on the response curve. So should the locations of the sound and adjacent-sound dips.

If the alignment was correct it should be possible to vary the bias and get essentially the same response curve. At higher bias values you will have to increase the sweep generator output and reduce the scope attenuation to compensate for the reduced gain of the i.f. amplifier.

In other respects i.f. sections for color or black-and-white reception have the same defects, such as lack of gain, regeneration and oscillation, and require the same cures, which include proper shielding, decoupling and grounding.

Alignment of Chroma Amplifier

In color TV receivers it may occasionally be necessary to realign the chroma amplifier. Normally this would only happen after repair work in the circuit. The chroma amplifier should have a bandpass of about 1 MHz with its center frequency at 3.579 MHz, the color subcarrier. Since the upper frequency limit is 4.1 MHz, which is very close to the 4.5 MHz sound i.f. frequency, 4.5 MHz traps may be included to eliminate any tendency for the signals to beat together, causing objectionable interference in the picture.

Alignment procedure should be given in the manufacturer's data, but if this is not available, proceed as follows:

1. Connect your test equipment as shown in Figure 8.4, except, do not connect the sweep and marker generators yet.
2. Turn everything on and let it warm up and stabilize.
3. Set the signal generator for an output of 4.5 MHz
4. Set the scope controls (see Figure 1.3) as follows:

SEC/DIV	2
TRIGGERING SOURCE	INT
TRIGGERING MODE	AUTO
VOLTS/DIV (CH.1)	.2
AC-GND-DC	AC

5. Set the bias box as for i.f. alignment.
6. Adjust each 4.5 MHz trap for minimum deflection on the scope.

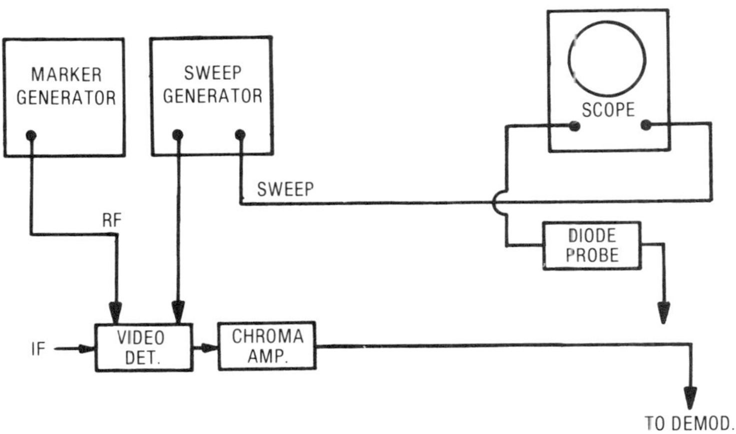

Figure 8.4 How to connect your test equipment to align the chroma amplifier.

7. Disconnect the signal generator.
8. Connect the sweep generator as shown in Figure 8.4.
9. Turn it on and let it warm up and stabilize.
10. Set the sweep generator for maximum sweep width and center frequency of 3.579 MHz.
11. Set the scope controls (see Figure 1.3) as follows:

X1-X10	X10
TRIGGERING MODE	X-Y
VOLTS/DIV (CH.1)	.1
AC-GND-DC	AC

12. Set a marker to 3.579 MHz, and others to 3.05 and 4.1 MHz, if available. If not, mark the positions of the 3.05 and 4.1 MHz markers on the scope screen with a grease pencil while the marker generator is set in turn to these frequencies. A response similar to that in Figure 8.5 should appear. If necessary, touch up tuned circuits for a flat-topped response between the 3.05 and 4.1 MHz markers.

Aligning the Tuner

Two distinctly different alignment tasks are necessary in the tuner. The oscillator frequency must be adjusted to obtain the correct i.f., and the tuner bandpass must be adjusted for the correct frequency response. Usually we do the bandpass first.

There are three TV bands, as follows:

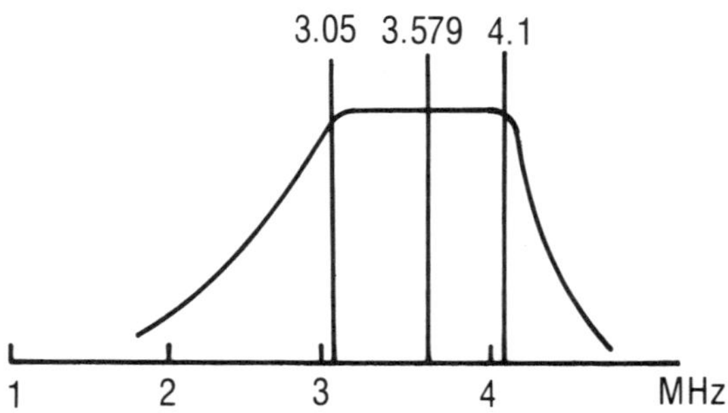

Figure 8.5 Ideal chroma response curve.

1. 54-72, 76-88 (Channels 2 through 6)
2. 174-216 MHz (Channels 7 through 13)
3. 470-806* MHz (Channels 14 through 69)

A standard RF sweep generator covers bands 1 and 2, but a UHF sweep generator is required for band 3.

Figure 8.6(a) shows the test equipment connections for tuner bandpass alignment. You'll notice that a matching network is needed to make the output of the sweep generator—50 or 75 ohms—compatible with the 300-ohm input to the tuner. If the tuner does not have a test point, connect to mixer grid (or transistor equivalent) via a 10-kilohm resistor.

The general procedure for tuner alignment is similar to that for i.f. bandpass alignment:

1. Turn everything on, and let it warm up and stabilize.
2. Set the sweep generator for maximum sweep width and center frequency of an unused channel.
3. Set the oscilloscope controls (Figure 1.3) as follows:

X1-X10	X10
TRIGGERING MODE	X-Y
VOLTS/DIV	2m
AC-GND-DC	AC

4. Adjust the bias box for the bias voltage specified by the manufacturer, or for 3 volts.
5. Check that there is no RF oscillation from the RF amplifier by reducing the sweep generator output to zero, and looking to see if there is any signal indication on both the scope and TV screens. If necessary, adjust the neutralizing capacitor.
6. Bring up the sweep generator output, and examine the response curve. It should look like that in Figure 8.6(b). It is not often you'll get one as flat and well proportioned, but it must resemble it as nearly as possible for good color reception. Carefully adjust the master trimmers and, if necessary, individual chan-

*Frequencies between 806 and 890 MHz, formerly allocated to television broadcasting, are now allocated to land mobile services.

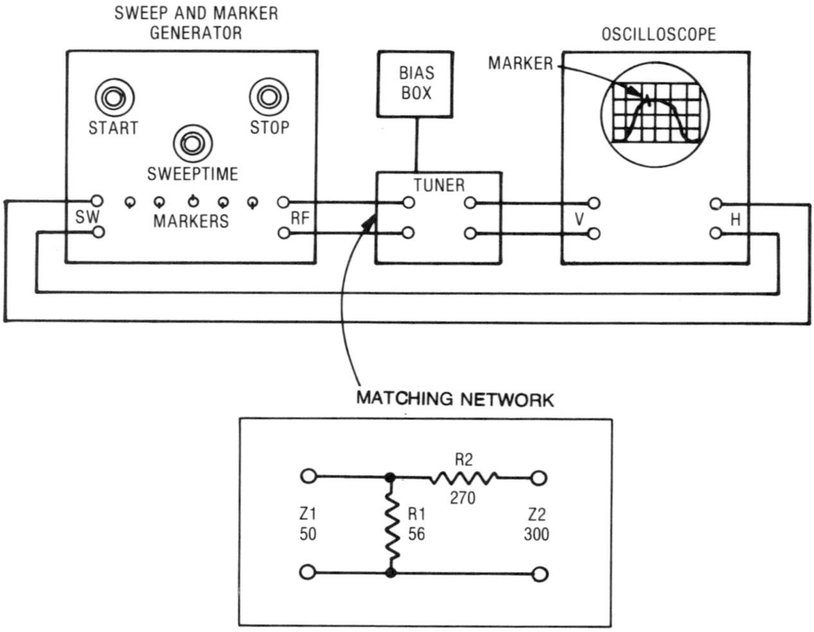

(a) HOW TO CONNECT YOUR TEST EQUIPMENT
FOR TUNER ALIGNMENT.

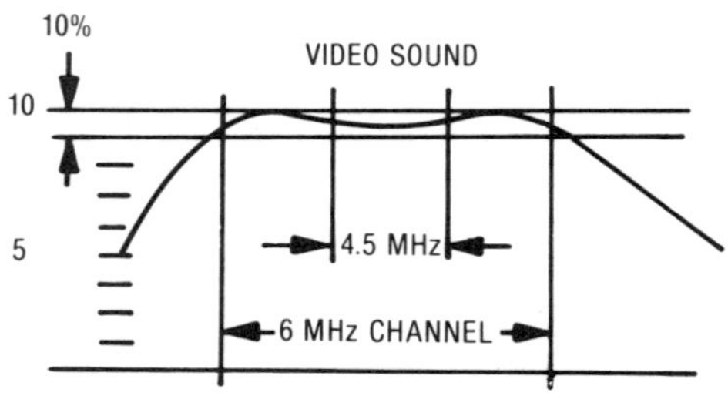

(b) IDEAL TUNER RESPONSE CURVE

Figure 8.6 Tuner alignment

nel coils to improve flatness. A deep "valley" in the middle of the response curve indicates too much coupling.

The local oscillator should be aligned after i.f. alignment, if that is to be done. It is simply a matter of adjusting each channel's oscillator tuning slug for the best picture on the TV screen. Set the fine tuning control to mid-range while doing this. The antenna should have been reconnected, of course.

Sound I.F. Alignment

In practically all color TV receivers the sound carrier is amplified along with the video in the i.f. section. In the video detector the 41.25 megahertz carrier is demodulated to become a new sound i.f. of 4.5 megahertz. This new signal has to be separated from the other components of the composite signal, the Y (or brightness signal) and the color subcarrier (or chroma signal). The latter has a frequency usually given as 3.58 megahertz (the exact figure is 3.579545 MHz) which is rather close to 4.5 megahertz, and will beat with it to give a 920-kilohertz difference signal if given the chance.

This 920-kilohertz signal will give an objectionable interference pattern on the TV screen if allowed to reach it, hence the use of 4.5-MHz traps in the video amplifier and chroma bandpass amplifier. The 41.25-MHz carrier is also kept at a level at least 50 decibels below that of the color subcarrier for the same reason.

In spite of these precautions, if the sound i.f. section is not tuned exactly to 4.5-MHz, so that it acts like a super trap, enough sound signal will reach the screen to spoil the picture. Therefore, the accurate tuning of the sound take-off coil and sound i.f. transformers is essential.

One method of aligning the entire sound section is to connect two signal generators, set at 41.25-MHz and 45.75-MHz, to the input of the mixer stage. The difference signal of 4.5-MHz will appear at the video detector. Connect your scope at the output of the sound i.f. section and adjust the sound trap and i.f. coils for maximum amplitude of the displayed signal.

Without touching anything else, reconnect the scope to the

audio output and adjust the ratio detector transformer for minimal deflection of the displayed signal. Then slightly vary the frequency of the signal from the 45.75 generator. Positive and negative deflections should be displayed as the generator frequency is shifted.

Another method of alignment is to apply an accurate 4.5-MHz signal to the input of the sound i.f. section. The ratio-detector transformer primary should be tuned for maximum output, and the secondary for zero. In the case of a gated-beam detector, the control-grid transformer is tuned for maximum output, while the quadrature grid coil is tuned for minimum.

FM Receiver Alignment

The alignment of an FM receiver consists of the following parts:

1. Aligning the tuner section
2. Aligning the i.f. section
3. Aligning the stereo-multiplex section (if stereo)
4. Aligning the audio section or sections

Always follow the manufacturer's instructions for best results. If these are not available the general procedure given below may be used.

Tuner Alignment

1. Connect signal generator to antenna terminals, using matching pad if required to obtain 300-ohm balanced input.
2. Connect scope to output of i.f. section. Use demodulator probe if scope bandwidth does not reach i.f. frequency (10.7 MHz). Set scope controls for maximum sensitivity and sweep speed to 0.1 or highest speed available.
3. Turn everything on, and allow sufficient time to warm up and stabilize.
4. Adjust signal generator for output of 98 MHz at about 100 μV, modulated 30 percent.
5. Set receiver tuning dial to 98 MHz.

6. Adjust slug in primary of tuner output transformer for maximum signal amplitude on scope screen.
7. Reduce signal generator output until noise starts to override signal.
8. Adjust trimmer capacitor of tuning capacitor for maximum signal amplitude on scope screen.
9. Reduce signal generator output as in 7.
10. Adjust slug in secondary of tuner output transformer for maximum amplitude as seen on scope screen.

I.F. Amplifier Alignment

1. Connect signal generator to input of i.f. amplifier.
2. Connect scope to output of i.f. section, as in 2. above.
3. Turn everything on, and allow sufficient time to warm up and stabilize.
4. Adjust signal generator for output of 10.7 MHz, at level that does not overload amplifier.
5. Adjust each i.f. transformer for maximum signal amplitude as seen on scope screen.
6. Reconnect scope to output of second detector (ratio detector, etc.), using compensated probe. Adjust detector transformer primary for maximum amplitude signal on scope screen, and secondary for minimum.
7. Rock signal generator tuning dial slightly each side of 10.7 MHz. Positive and negative deflections should be displayed as generator frequency is shifted.

Aligning the Stereo-Multiplex Section

1. Connect signal generator to stereo-multiplexer input.
2. Connect scope to output of 38-kHz amplifier.
3. Turn everything on, and allow sufficient time to warm up and stabilize.
4. Adjust signal generator for output of 19 kHz. Use minimum signal to avoid overloading circuits.
5. Touch up i.f. transformer adjustments as in i.f. alignment, for maximum signal amplitude on scope.
6. Transfer scope probe to output of 19-MHz trap, and adjust for minimum signal amplitude on scope.

Aligning Audio Section

There really isn't anything to adjust in the audio section. You should, however, check performance as explained in Chapter 6.

Aligning a Ratio Detector

Many FM receivers and television sound sections use a ratio detector similar to that in Figure 8.7 (a). To check its response, proceed as follows:

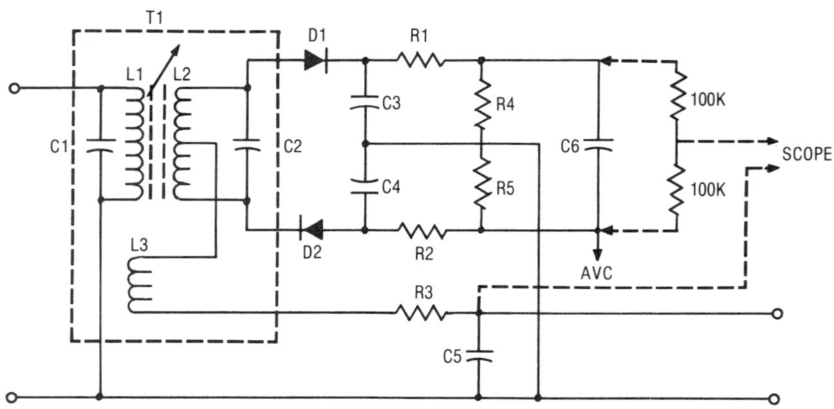

Typical ratio detectors circuit, showing where to connect resistors and scope.

Figure 8.7 (a) Aligning a Ratio Detector

1. Join together two 100-kilohm resistors in series, and then connect them in parallel with C6.
2. Connect the scope vertical input terminal to the junction of the two 100-kilohm resistors, with the scope ground return connected to the junction of R3 and C5.
3. Connect your sweep and marker generator to the input of the sound i.f. amplifier. Connect the sweep output to the horizontal input of the scope as in Figure 8.1.
4. Turn everything on, and allow sufficient time for the equipment to warm up and stabilize.
5. Adjust the sweep and marker generator for a center frequency of 10.7 or 4.5 MHz, with a sweep of 1 MHz, and a 10.7 or 4.5 MHz marker, as the case may be (FM or TV).

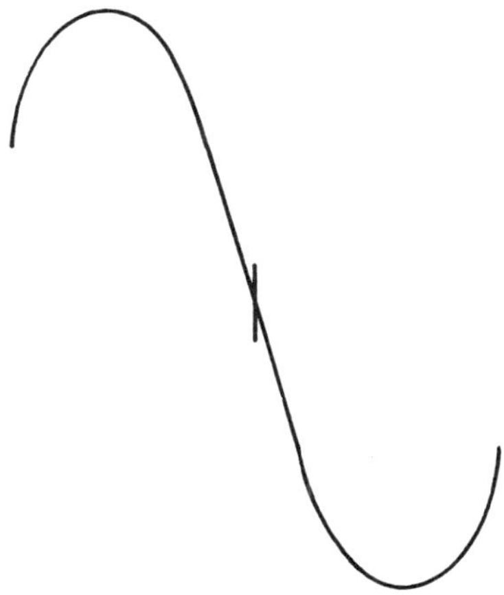

Figure 8.7 (b) Typical S-curve response you should see on your scope.

6. Your scope screen should display a response similar to Figure 8.7 (b). The S curve should be symmetrical about the marker. If it is not, adjust T1's secondary as necessary.
7. When you've finished, be sure to remove the two 100-kilohm resistors connected across C6.

9

Waveform Analysis

Analyzing a waveform means:

a. Comparing the displayed wave shape with the ideal shape.

b. Measuring the amplitude (in volts).

c. Measuring the frequency (in hertz).

d. Measuring the width, risetime, interval (in seconds).

All complex waveforms can be analyzed in terms of sine-wave components, so we'll start with *sine waves*. Their name goes back to the early alternating generator of Elihu Thomson in 1878. In this original two-pole a.c. generator the output voltage or current rises smoothly from zero to a maximum positive value, decreases to zero, falls to a maximum negative value, and returns to zero, during each revolution of the rotor. A complete revolution is 360 degrees, of course, and at each instant the value of the output voltage or current is proportional to the *sine* of the angle through which the rotor has turned up to that instant. Any wave having this form is called a sine wave, whatever its source. If it is approximately of this form it is called a *sinusoidal wave*.

A sine wave is illustrated in Figure 9.1, which also gives the various terms used for its dimensions. Many of these terms are also used for other waves.

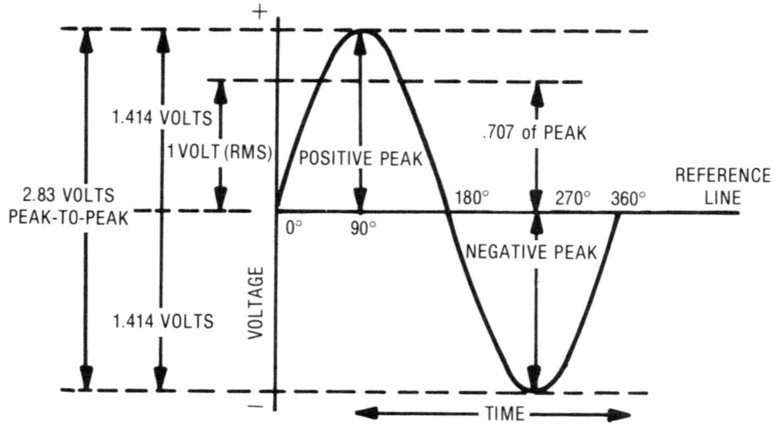

Figure 9.1 The dimensions of a sine wave. The amplitude at any point is proportionate to the sine of the "angle" measured along the reference line from 0° to the point.

Modulated and Mixed Waveforms

Modulated and mixed waveforms both contain two (or more) frequencies. However, in an amplitude-modulated wave the amplitude of the carrier is modulated, as shown in Figure 9.2 (A), with the production of sideband frequencies, which are new frequencies that were not present in the generator of the carrier or in the generator of the modulating signal. This type of mixing is done in a non-linear mixer or modulator.

The mixer in, say, an audio unit is not a signal processor, and is a linear mixer. No sidebands are formed, and the resulting waveform contains the original frequencies only, as in Figure 9.2 (B). Another example of linear mixing is seen in the NTSC color-bar signal, as shown in Figure 9.3 (A), which is made up of the Y signal and the color subcarrier, (B) and (C).

Wave Dimensions and Terms

Figure 9.4 shows two complete cycles (AE and EI) of a continuous, undamped sine wave. The conventional dimensions of the

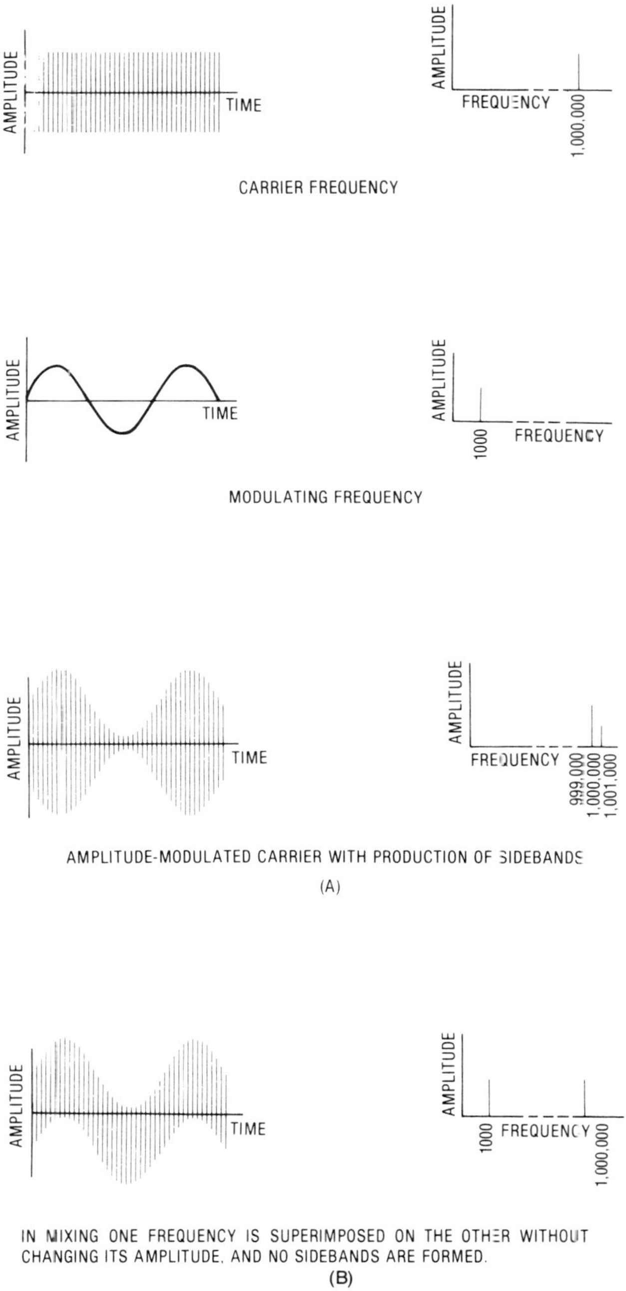

Figure 9.2 Modulation and mixing

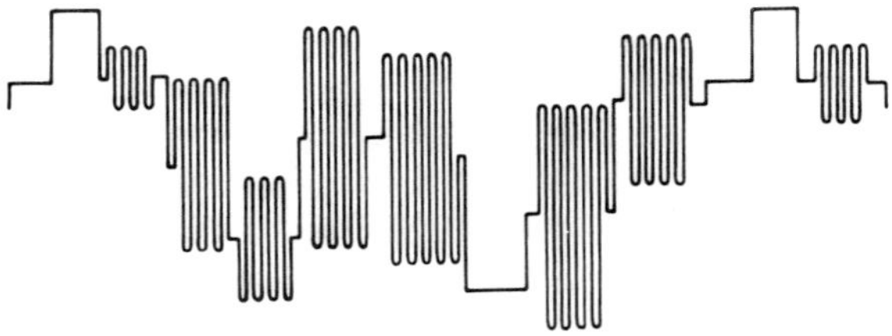

(A) COMPLETE COLOR-BAR SIGNAL

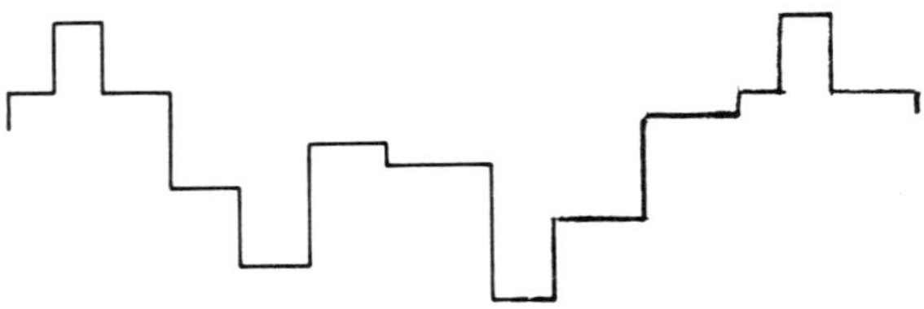

(B) Y-SIGNAL COMPONENT OF (A)

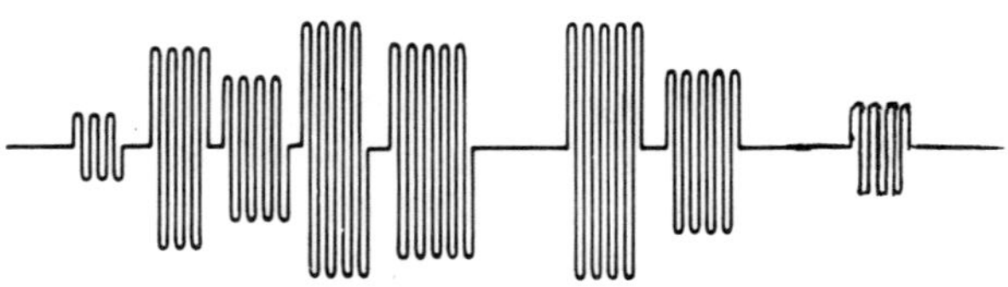

(C) COLOR SUB-CARRIER COMPONENT OF (A)

Figure 9.3 Another example of mixed waveforms

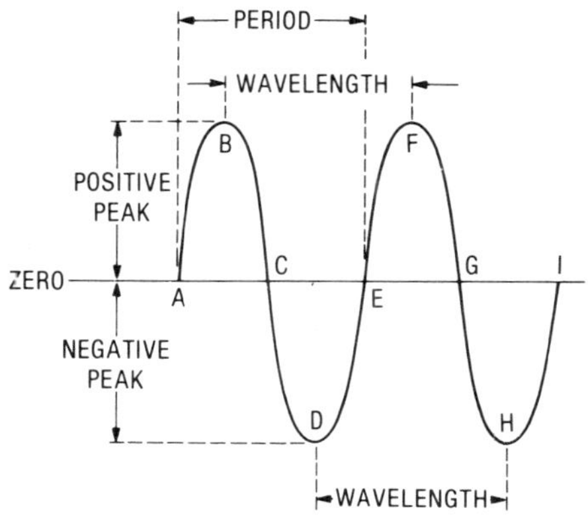

Figure 9.4 Conventional terms used to dencte wave characteristics

wave, which are indispensable in discussion and measurements, are described below.

Zero Line. This horizontal line represents zero current or voltage. Thus, the current or voltage of the wave is zero at points A,C,E,G, and I.

Positive Peak. The maximum positive (or positive-going) amplitude which the current or voltage reaches in the wave pattern. It is a peak value (i.e., 1.414 rms or 1.572 avg) designated +E for voltage or +I for current. The wave shown is at its positive peak at points B and F.

Negative Peak. The maximum negative (or negative-going) amplitude which the current or voltage reaches in the wave pattern. It is a peak value (i.e., 1.414 rms or 1.572 avg) designated −E for voltage or −I for current. The wave is at its negative peak at points D and H.

Period. The time (in seconds) required for the completion of one full cycle of the wave. Since the wave shown starts at zero and goes through one positive and one negative peak before ending at zero, its period is the horizontal dimension AE. The horizontal axis represents time. Since two cycles are shown here, the period can also be measured between E and I. Not all wave patterns start and stop at zero; some start and stop at some instantaneous value of current or voltage. But the wave must stop at the same value at which it started, and the period is measured between those two points. The symbol for period is T.

Frequency. The number of cycles completed by the current or voltage wave in a single second of time. (Two cycles, ABCDE and EFGHI, are shown in Fig. 9.4) A familiar example is common 60 hertz a.c. power, which goes through 60 cycles each second. The symbol for frequency is f and for hertz, Hz. The use of very high frequencies in electronics makes the expression of some frequencies in hertz (Hz) unwieldy, so kilohertz (1 kHz = 1000 Hz), megahertz (1 MHz = 1000 kHz), and gigahertz (1 GHz = 1000 MHz) are used, as required.

Wavelength. Conventionally, the length of a wave is measured from one peak to the next adjacent one; i.e., from B to F, or from D to H, in Fig. 9.4. Wavelength is expressed in meters, and its symbol is λ (lambda).

Angular Velocity. When the rotor of an electromechanical a.c. generator turns through 360°, completing one full revolution, it describes 2π radians (1 radian = approximately 57.3°), and the angular velocity of the shaft is 2π radians multiplied by the speed of the rotor in revolutions per second. A single cycle of the alternating current or voltage output of this machine, or of a comparable electronic generator, is completed in 2π radians, and its frequency (f, in Hz) is analogous to the speed of the shaft in rps. Hence, angular velocity is expressed electrically as 2π times the a.c. frequency. The symbol for angular velocity is ω.

$$\omega = 2\pi f \quad \text{where } \pi = 3.1416 \qquad \textbf{(9.1)}$$
$$f = \text{frequency in Hz.}$$

Table 9.1 shows angular velocity values for some common frequencies.

Wavelength-Period-Frequency Relationships. Since wavelength, period, and frequency of a given wave are mutually related, one quantity may be found in terms of another. The relationships are shown below.

WAVELENGTH

$$\text{meters} = \frac{300 \times 10^6}{\text{fHz}} = \frac{300{,}000}{\text{fkHz}} = \frac{300}{\text{fMHz}} = \frac{0.3}{\text{fGHz}} \qquad \textbf{(9.2)}$$

$$\text{meters} = 300 \times 10^{6}\text{Ts} = 300{,}000\text{Tms} \quad \textbf{(9.3)}$$
$$= 300\text{T}\mu\text{s}$$

FREQUENCY

$$\text{fHz} = 1/\text{Ts} = 1000/\text{Tms} = 10^{6}/\text{T}\mu\text{s} \quad \textbf{(9.4)}$$

$$\text{fkHz} = 0.001/\text{Ts} = 1/\text{Tms} = 1000/\text{T}\mu\text{s} \quad \textbf{(9.5)}$$

$$\text{fMHz} = 10^{-6}/\text{Ts} = 10^{-3}/\text{Tms} = 1/\text{T}\mu\text{s} \quad \textbf{(9.6)}$$

PERIOD

$$\text{Ts} = 1/\text{fHz} = 10^{-3}/\text{fkHz} = 10^{-6}/\text{fMHz} \quad \textbf{(9.7)}$$
$$= \lambda/(300 \times 10^{6})$$

$$\text{Tms} = 10^{3}/\text{fHz} = 1/\text{fkHz} = 10^{6}/\text{fMHz} \quad \textbf{(9.8)}$$
$$= \lambda/300{,}000$$

$$\text{T}\mu\text{s} = 10^{6}/\text{fHz} = 10^{3}/\text{fkHz} = 1/\text{fMHz} = \lambda/300 \quad \textbf{(9.9)}$$

Frequency (f)	Angular Velocity (ω)
20 Hz	125.66
50	314.16
60	376.99
100	628.32
400	2513.28
500	3141.60
1000	6283.20
100 kHz	628,320
1 MHz	6.2832×10^{6}
10	6.2832×10^{7}
100	6.2832×10^{8}

Table 9.1. Angular velocity v. frequency

Alternating Waves

Alternating waves rise once to a positive peak and then fall once to a negative peak in each single cycle. Common alternating waves are shown in Figure 9.5.

Note that in each instance current or voltage rises from zero to a positive peak amplitude, returns to zero, falls to a negative peak amplitude, and finally returns to zero. The cycle then may be repeated; in fact the wave continues as long as the generator is in operation. A phenomenon is alternating only if it is periodically negative and positive.

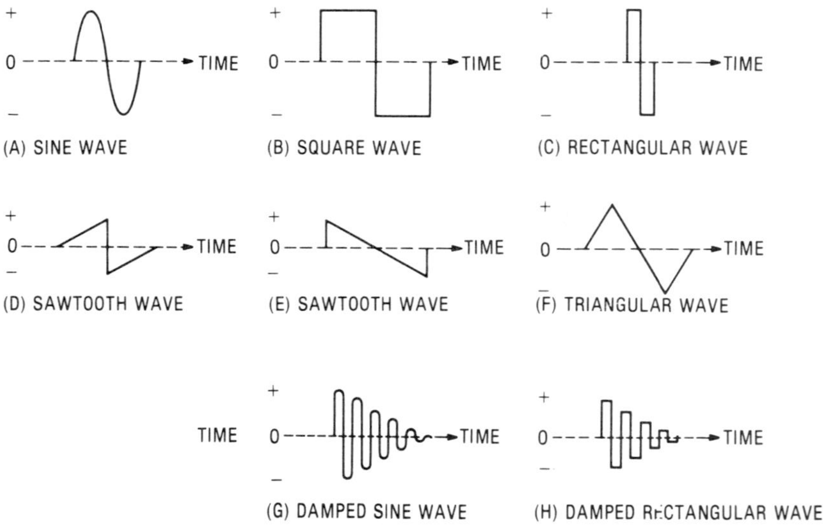

Figure 9.5 Common types of alternating waves

Pulsating Waves

Pulsating waves rise from zero to a positive peak amplitude and then return to zero, or fall from zero to a negative peak amplitude and return to zero. Unlike an alternating wave, therefore, a pulsating wave is always of one polarity even though rising or falling in value, as in Figure 9.6.

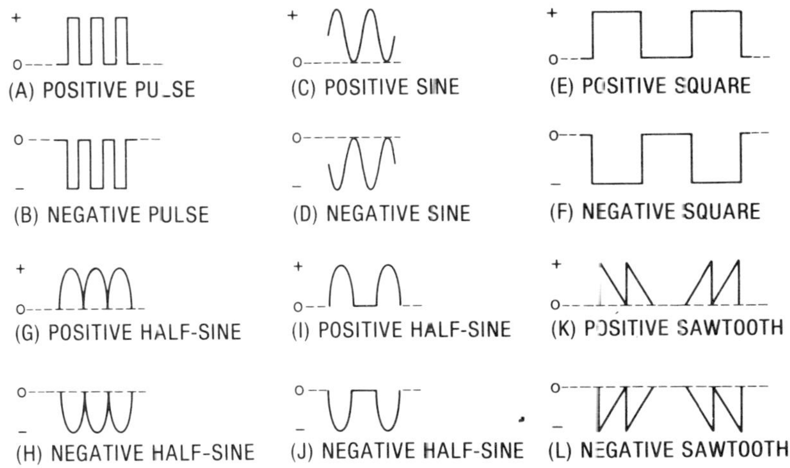

Figure 9.6 Common types of pulsating waves

Pulse Characteristics

Pulses belong in the *time domain,* as opposed to sinusoidal waves, which occupy the *frequency domain.* In both domains the amplitude of the signal is an important parameter. But in the time domain the other parameters are mostly concerned with time intervals, as we shall see in a minute.

Pulse signals have many uses. For example, they are used in:

Radar and navigational equipment
Communications (pulse modulation)
Telemetry
Computers
Process control

The *baseline* (OT in Figure 9.7) is the reference line for a pulse. It may be at zero potential, or at a positive or negative potential. In the latter case it is said to be *offset* by the value of the d.c. potential. Dimensions normal to the baseline represent amplitude, those parallel to it represent time. The time during which any portion of the pulse waveform coincides with the baseline (zero amplitude) is the *baseline dwell time*. Pulses may make excursions on either side of the baseline, but must have baseline dwell times between excursions.

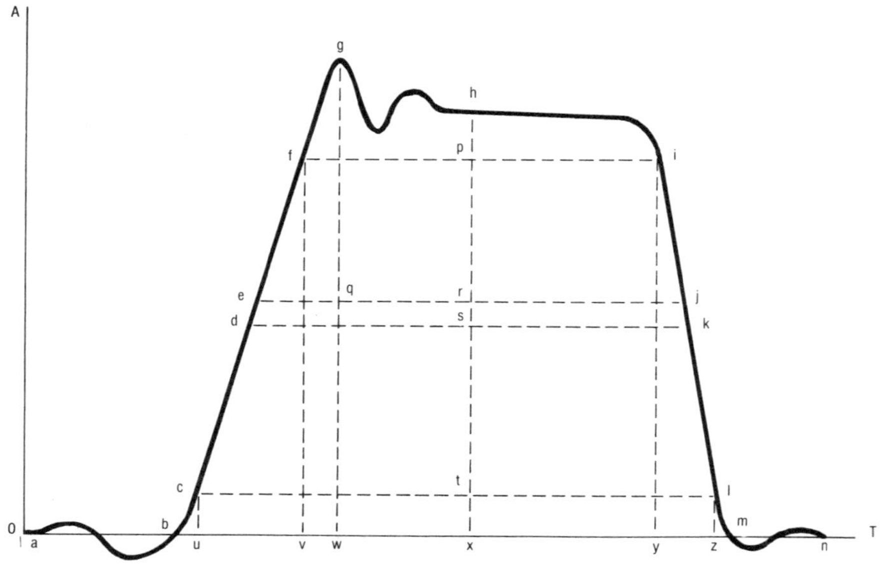

Figure 9.7 Pulse geometry

gw	= peak amplitude
e	= leading edge peak median
j	= trailing edge peak median
h	= centrum: er = rj
hx	= pulse amplitude; hs = sx
d	= leading edge median
k	= trailing edge median
dk	= pulse width
c	= leading edge decium ("10% point"): cu = 10% of hx
f	= leading edge nonum ("90% point"): fv = 90% of hx
uv	= risetime
i	= trailing edge nonum
l	= trailing edge decium
yz	= fall time
abcd	= initial corner
defgh	= leading edge corner
hijk	= trailing edge corner
klmn	= terminal corner

A *trapezoidal pulse* has a flat top parallel to the baseline, with a leading edge that slopes up from the baseline to the top, and a trailing edge that slopes downward from the top to the baseline, as in Figure 9.6. As the leading and trailing edges are made steeper the pulse approaches the shape of a rectangle. It can never become a perfect rectangle because to do so the amplitude would have to rise from the baseline to the top in zero time, which is not possible. Some pulses look on an oscilloscope screen as if they do just that, but if the sweep rate is increased enough you can see that it is only because of the terrific speed of their rise and fall times that such pulses *seem* to be rectangular.

Rectangular pulses generally have a form resembling that in Figure 9.7. A very narrow, steep-sided pulse is a *spiked* pulse, while one where the corners are smoothly rounded is a *rounded* pulse.

The pulse dimensions shown in Figure 9.7 are based on the *peak amplitude,* which is the maximum amplitude that can be measured vertically from the baseline to any point on the pulse. You should not confuse this with the *pulse amplitude,* which is the amplitude of the *centrum* above the baseline.

The *rise* and *fall times* of the pulse indicate its steepness. The steeper the risetime, the more likely is the leading edge to overshoot and cause a ripple in the first part of the pulse top. Such an overshoot and ripple (called *ringing*) are known as *pulse top distortion.* Since many amplifiers will cause similar distortion to an otherwise good pulse you need to know what the pulse looked like before it was applied to the amplifier, for comparison.

The pulse equivalent of frequency (number of pulses per second) is the *pulse repetition rate* (PRR): it is wrong to speak of "pulse frequency" because we are not in the frequency domain. Pulses, lacking the unbroken form of a continuous wave, have much more the nature of chopped DC, and to attempt to apply frequency-domain mathematics to them doesn't work too well. The exception is the square wave, which is a continuous wave having many pulse characteristics also.

Fluctuating Waves

Like pulsating waves, *fluctuating waves* (also called *composite* waves) are confined to one side of the zero current or voltage axis. But, unlike pulsating waves, they never touch zero. Thus, a fluctuating wave may rise from an initial current or voltage level to a peak amplitude and return to the initial level, or it may swing cyclically to a peak amplitude on each side of the initial level just as an alternating wave swings on each side of zero. The chief difference is the d.c. content of the fluctuating wave. Because of this mixture of a.c. and d.c. the fluctuating wave is often called *composite*.

Typical of fluctuations, each of the waves shown in Figure 9.8 lies wholly in the positive region or wholly in the negative. In Figure 9.8(a), for example, the upper pattern consists of alternate peaks ($+e_3$ and $+e_1$) above and below a reference voltage ($+e_2$), all in the positive region. Similarly, the lower pattern consists of alternate peaks ($-e_3$ and $-e_1$) above and below a reference voltage ($-e_2$), all in the negative region. (Corresponding current values would be $+i_3$, $+i_1$, $+i_2$; $-i_3$, $-i_1$, $-i_2$.) An alternating or pulsating

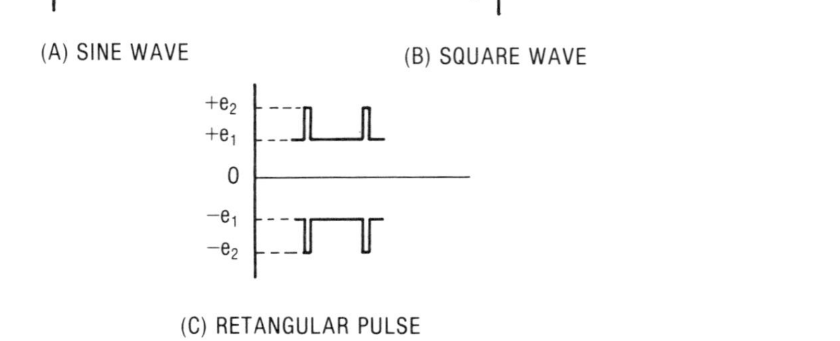

Figure 9.8 Fluctuating (or composite) waves "ride on d.c."

wave of any shape may be superimposed upon a d.c. component in this manner to produce a fluctuating wave. Fluctuations are common in the electrode currents and voltages of tubes and transistors handling a.c. signals.

When the wave has an "alternating" shape, the various voltages are designated by the following terms: e_1, *trough voltage*; e_2, *mean voltage*; e_3 *peak* (or *crest*) *voltage*. Currents are similarly designated, using *i* symbols. Fluctuations on the positive side of zero are said to be *positive-going* when they increase in the direction away from zero, and *negative-going* when they decrease toward zero. Conversely, fluctuations on the negative side of zero are said to be *negative-going* when they increase in the direction away from zero, and *positive-going* when they decrease toward zero. The term *positive-going* thus signifies becoming more positive (or less negative), whereas *negative-going* signifies becoming more negative (or less positive).

Symmetry and Asymmetry

A wave is *symmetrical* when one half of it is a mirror image of the other half. Thus, the negative half-cycles of a symmetrical a.c. wave are identical with the positive half-cycles. A wave is asymmetrical (or unsymmetrical) when it lacks this regularity, as in Figure 9.9.

Fundamental and Harmonic Frequencies

The *fundamental frequency* of a wave is that frequency indicated simply by the number of cycles per second. But no practical

(a) (b)

Figure 9.9 Symmetrical wave (a) and asymmetrical wave (b).

generator produces so pure a wave; undesired higher frequencies are generated simultaneously with the fundamental. These higher frequencies, termed *harmonics,* are integral multiples of the fundamental and thus are comparable to the overtones of a sound. Harmonics are numbered according to their numerical relationship to the fundamental: 2nd harmonic (twice fundamental), 3rd harmonic (three times fundamental), etc. Technically, the fundamental is the 1st harmonic but it is rarely called by that name.

All practical electric wave signals are complex in the sense that they are composed of a fundamental and one or more harmonics. A close analysis of nonsinusoidal waves, such as the square, sawtooth, and triangular types, shows that they may be considered also as consisting of a number of sine waves (the fundamental and various harmonics). These various frequencies interact, and the resultant wave has the nonsinusoidal shape. The more frequencies there are in the combination, the more square-, rectangular-, or sawtooth-shaped is the wave.

The Meaning of Distortion

When a waveform departs from its ideal shape, it is said to be distorted. The signals shown in Figure 9.9 are distorted since asymmetry causes them to differ from the ideal sine wave. In sine waves *harmonic distortion* results from the presence of harmonics; the higher the ratio of harmonic energy to fundamental energy, the more distorted is the wave shape. Such distortion may be caused by a number of factors, such as nonlinear operation of components or circuits.

Figure 9.10(a) shows distortion resulting from a high percentage of 3rd harmonic energy in a wave. The manner in which the fundamental and 3rd harmonic interact to produce this shape is illustrated by Figure 9.10(b). Here, the fundamental and harmonic have various instantaneous amplitudes with respect to each other, and sometimes the same and sometimes the opposite polarity. The result is that the harmonic adds to the fundamental at some instants and subtracts from it at others, the resultant of this action being the distorted wave. From this example, it should be clear that the shape of a distorted wave depends upon which and how many harmonics are present and upon their strengths and phase relationships as well.

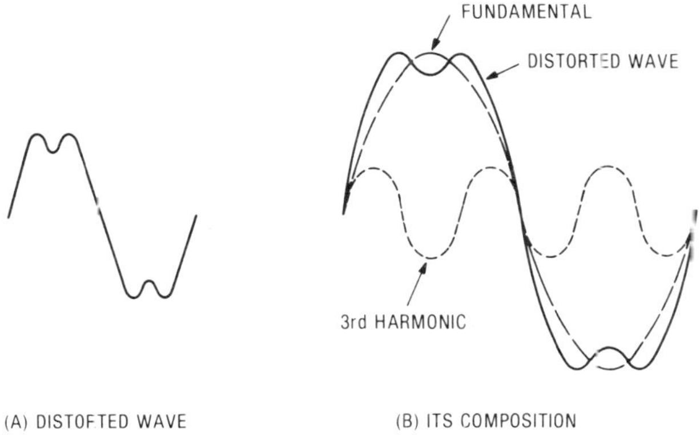

Figure 9.10 How harmonic distortion is caused

Very small distortion percentages make so little a change in the shape of a wave that it is detectable only by very close inspection of the shape.

Simple Observation of Waveform

Uncritical demands are often satisfied by a simple visual inspection of a wave pattern, with few, if any, quantitative measurements. For example, you may see a pattern on your scope screen and note that it is not a good square wave, or you may find ringing in the horizontal deflection circuit of a TV receiver (the frequency and damping of the disturbance are relatively unimportant here, the mere presence of ringing indicating trouble).

The scope is undoubtedly used more often than any other instrument for the simple observation of waveforms. A glance at the displayed pattern shows quickly the correspondence of the signal to the desired wave shape. It will also show relative amplitude and phase. If this visual appraisal is insufficient, quantitative values may be obtained by measuring the pattern's dimensions.

Figure 9.11 shows general test setups. To check waveform of an a-f oscillator, connect the vertical amplifier input terminals of the scope directly to the output terminals of the oscillator, as shown in (A). Then, reconnect the vertical amplifier input terminals to a suitable load device which, in turn, is connected to the output terminals of the amplifier, as shown in (b). For quiet checking of a power

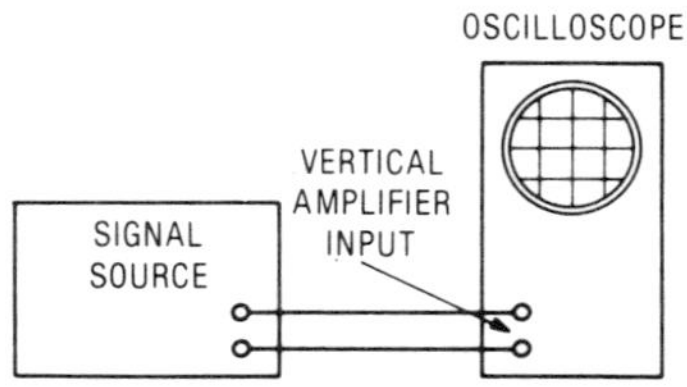

(A) CHECKING SIGNAL SOURCE

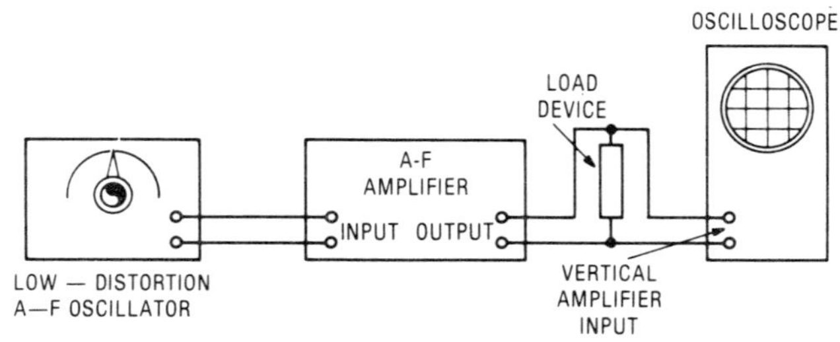

(B) CHECKING OUTPUT OF AMPLIFIER

Figure 9.11 How to set up your test equipment to examine output waveforms of audio amplifiers.

amplifier, the load device is a noninductive resistor equal in resistance to the output impedance of the amplifier and rated at twice the power output of the amplifier.

PROCEDURE

1. Connect the VERTICAL INPUT terminals of the scope to the signal-source terminals.
2. Set scope TRIGGERING SOURCE to INTERNAL and MODE to AUTO.
3. Switch on scope and allow for warmup.
4. Switch on signal source.
5. Adjust SEC/DIV and TRIGGERING controls for a single, stationary signal cycle on the screen.
6. Adjust VOLTS/DIV control for desired width and height of pattern.
7. Adjust FOCUS and INTENSITY controls for sharp pattern.
8. Reconnect vertical input terminals of scope to amplifier output and readjust VOLTS/DIV control.
9. Compare pattern wave shape to ideal shape, i.e., sinusoidal, square, triangular, etc. (See Figure 9.5.)

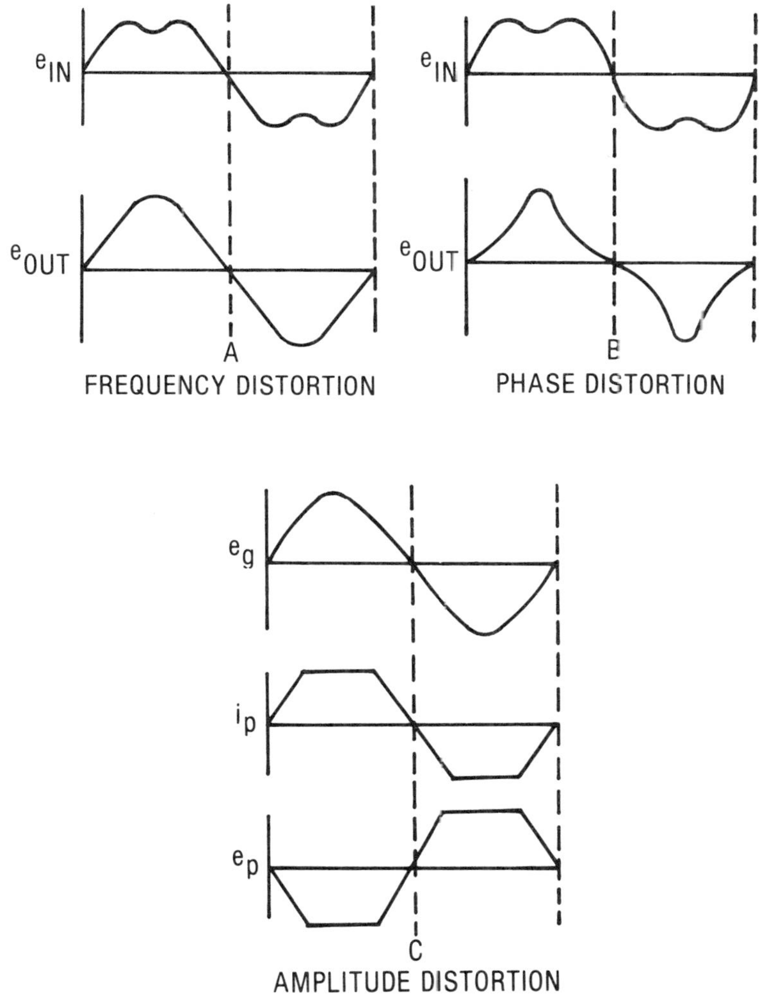

Figure 9.12 Types of amplifier distortion

The following pointers are important: pulsating waves (Figure 9.6), fluctuating waves (Figure 9.8), and asymmetrical waves (Figure 9.9) may be satisfactorily analyzed only if you know where the zero axis or d.c. axis actually is, and this can be known only when the scope AC-GND-DC input switch is set to DC. Low distortion percentage (e.g., less than 5 percent total) usually cannot be detected by sight alone; a low-distortion pattern could pass for a perfect sine wave unless measurements are made.

A technician who regularly uses his scope to make rapid spot checks of waveform looks for certain faults. Some of these are (1) kinks, dimples, flattening, or rippling of sinusoidal waves; (2) rounded corners or blunt points in square, rectangular, sawtooth, or triangular waves; (3) vertical tilt, horizontal slope, curved or crooked lines, spikes, and ringing in square and rectangular waves; and (4) nonlinearity in sawtooth and triangular waves.

Distortion is any undesirable change in the waveform between the input and the output of the amplifier. Three fundamental types of distortion are amplitude, frequency and phase distortion (see Figure 9.12).

Amplitude distortion is what happens when the amplitude of the output signal is not a linear function of the input amplitude. This means that the output waveform does not have the same form as the input wave. For example, when a strong sine-wave signal is clipped as shown in Figure 9.12(c) because the operating point of one of the amplifier stages has changed we have amplitude distortion.

Frequency distortion occurs when the amplifier fails to give equal amplification to all audio frequencies. In Figure 9.12(a) the input signal to an amplifier contains the fundamental and third harmonic, but only the fundamental appears in the output.

Phase distortion is where the phase of the output signal is displaced relative to the input. If all frequencies were displaced by the same phase angle, it wouldn't matter, but they are not. Consequently, the phase angles between signals of different frequency are altered as they pass through the amplifier, resulting in alteration of the waveform. This is illustrated in Figure 9.12 (b), where the third harmonic has been phase-shifted more than the fundamental, completely altering the waveform.

Harmonic distortion is produced when harmonics of fundamental frequencies are generated as the signal passes through the amplifier system. Harmonics are weaker than fundamentals, of course, but they produce severe distortion as shown in Figure 9.13. Harmonic distortion always accompanies amplitude distortion.

Intermodulation (IM) distortion is undesired amplitude modulation of one signal by another, arising from the non-linear response of the amplifier. Since both signals are audio frequencies it is not true amplitude modulation. You would test the amplifier by applying two signals to it simultaneously, with frequencies of, say

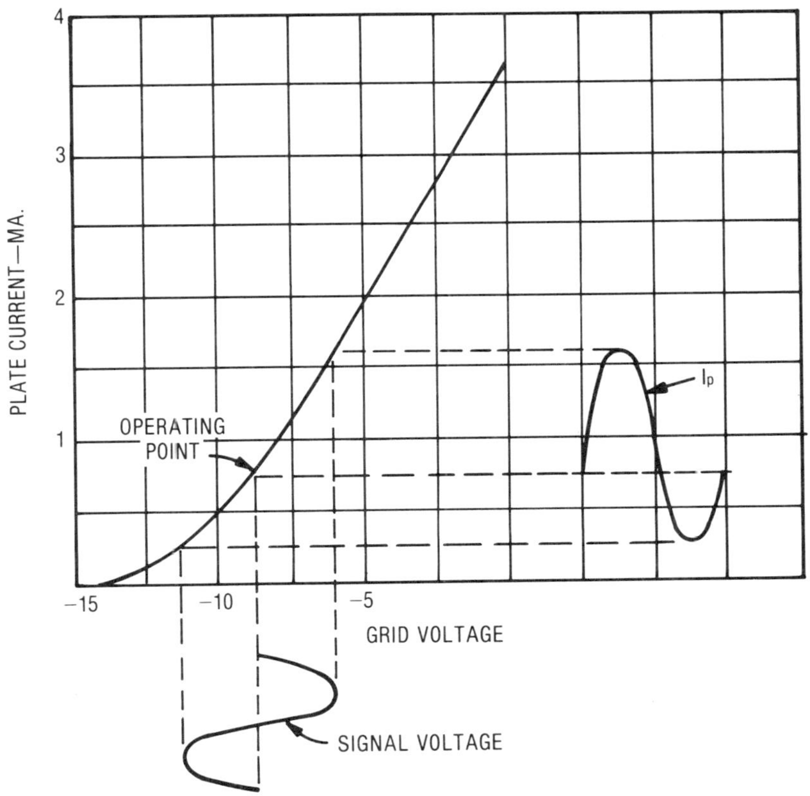

Figure 9.13 An example of how harmonic distortion may arise. An incorrect voltage on the grid of an amplifier tube places the signal on the curved portion of the curve instead of the straight position, so that the output current (Ip) is non-linear (asymmetrical). The same thing could happen to a transistor amplifier.

60 Hz and 7000 Hz. If IM is not present, your scope will show only the 7000-Hz signal with a wavy border at the 60-Hz rate.

Since most audio signals of speech and music, for example, are composed of fundamentals and many harmonics (Figure 9.10), you can see how garbled they can become in a poor amplifier as the original signals are distorted in their various ways. Signals with many harmonics, such as those produced by the sharp sounds of percussion or plucked strings, have waveforms with steeply rising fronts that may also shock some circuits into oscillation. The oscillation is usually a damped oscillation that adds to the signal fed into the amplifier, and in some cases will cause a very definite ringing sound in the output. Transient oscillations like these that die away in a few cycles may be caused by feedback loops acting as RC oscillators, or by self-resonant transformer or loudspeaker windings.

In an audio amplifier, continuous oscillation at a frequency above the range of hearing may be caused by a defective feedback network or decoupling capacitor. Although inaudible themselves, these oscillations may nevertheless overload the amplifier, causing amplitude distortion in the desired signal. In some cases they may cause IM distortion by beating with the desired signal, to give spurious audible tones in the output. This type of oscillation may only break out with a strong signal, because there is insufficient positive feedback from a weak signal to excite it. It is visible on the oscilloscope as a modulation envelope on the audio waveform.

As mentioned already, a square wave may be considered as a fundamental sine-wave frequency with an infinite series of odd harmonics. (Figure 9.14.) Thus the square wave will appear distorted at the output of the amplifier if the amplifier is unable to give all of its frequencies equal treatment. A one-kilohertz square wave will show how the amplifier handles them from that frequency to the limit of human hearing; a 50-hertz square wave will do the same for the lower frequencies. If an audio amplifier can reproduce square waves of both frequencies satisfactorily you can consider it to be free of frequency and phase distortion over the entire audio spectrum.

The higher-order harmonics determine the shape of the rising and falling edges of the square wave. If there is loss of the higher frequencies, those portions become rounded. The lower frequencies determine the shape of the horizontal portion of the square wave, so that it sags if they are attenuated.

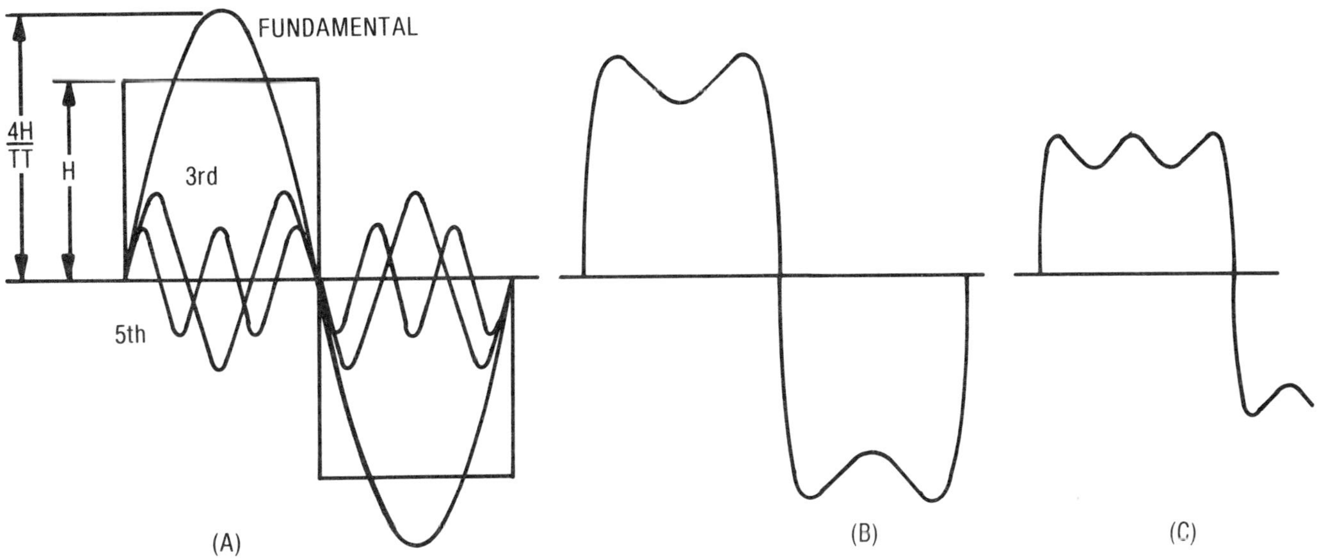

Figure 9.14 Composition of a square wave, showing the first three harmonics, (A) and the approximation to a square wave as the first two, (B) and the first three, (C) odd-harmonics are combined.

The output of your square-wave generator should never be high enough to overdrive any of the amplifier's stages, and it is essential to use an oscilloscope that does not itself distort the signal.

Using Lissajous Figures to Check for Distortion or Phase Shift

We've mentioned Lissajous figures elsewhere (Chapters 1 and 5, for instance) as being patterns produced on the scope screen when sine-wave signal voltages are applied to the Y and X inputs at the same time. This is a good method to use when checking any type of audio amplifier for distortion or phase shift.

Hook up the amplifier and test equipment as shown in Figure 9.15. The audio oscillator is connected to the input of the amplifier and to the Y (vertical) input of the scope. The output of the amplifier is applied to the resistor R, which should have the same resistance value as the load the amplifier normally works into, and sufficient wattage for the power developed. The X (horizontal) input of the scope should be connected across this resistor.

Turn the equipment on, and adjust the audio oscillator output for a 1-kilohertz signal, with an amplitude high enough to allow the audio amplifier to develop its maximum rated power output. Adjust the Y and X amplifiers of the scope for a Lissajous figure that slopes at an angle of 45 degrees, and is of a size that will fit on the screen. The slope will be from left to right upward if the amplifier's output has the same polarity as its input, or downward if it is inverted (180 degrees phase difference). Assuming it is inverted with respect to the input, you will get one of the patterns shown in figure 9.16.

Pattern (a) is the normal display, indicating no phase shift or overload. Moderate overload, without phase shift, will give pattern (b), bent at one end; and severe overload, without phase shift, results in pattern (c). When the pattern is an ellipse, as in (d), phase shift is indicated. Some phase shift occurs in many amplifiers near the high or low-frequency cutoff level, but it should be negligible at 1 kilohertz. Phase shift patterns with moderate and severe overload are shown at (e) and (f).

Two other types of distortion without phase shift are shown at (g) and (h). In (g) the diagonal line of (a) is curved. Incorrect bias voltage on a tube or transistor will cause this, by shifting the operat-

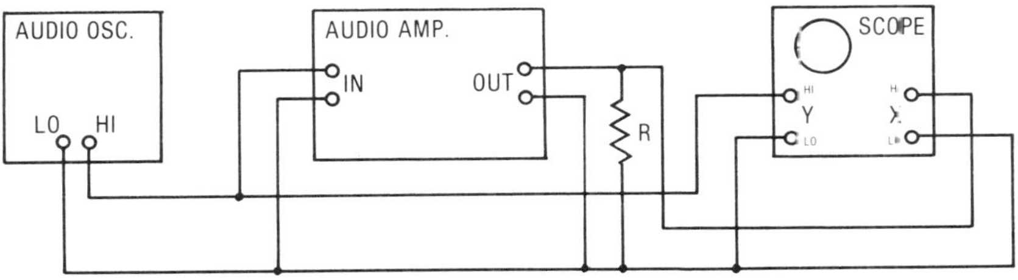

Figure 9.15 Hookup for Lissajous Figure Testing

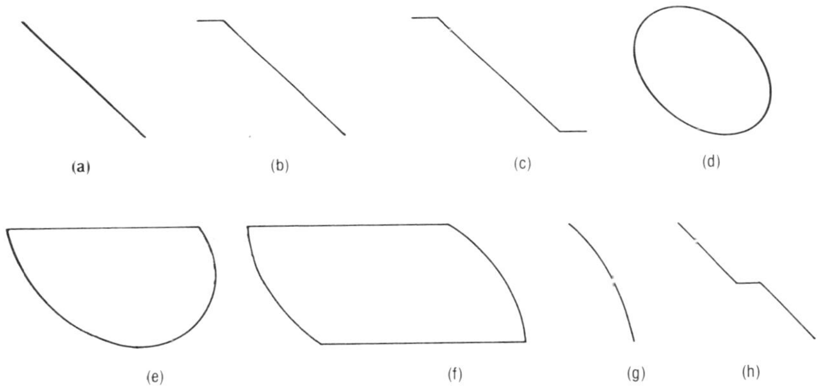

Figure 9.16 Lissajous patterns that may appear

ing point into the non-linear portion of the characteristic curve. In (h) we see the effect of incorrect bias in a push-pull stage. This is known as crossover distortion.

After performing the check with a 1-kilohm signal, it should be repeated at several other frequencies over the frequency range of the amplifier.

Square-wave testing of the amplifier has already been covered, so we need not go over it again. The service literature for a hi-fi amplifier will often show how a square wave of some specified frequency and amplitude should look.

You can also use your scope to perform a separation test of a stereo amplifier system. For this you need a stereo signal generator. When you apply a right signal your scope should show a minimum of 10 decibels between the signal at the right output and that at the left, and vice versa. A good multiplex unit should be capable of about 30 dB of separation.

10

Calibration and Maintenance

In this chapter we shall explain how to keep your scope in good working order, and how to make adjustments and repairs when necessary. Caring for your scope is like caring for your automobile. It's better to provide routine maintenance than wait for a break down.

How often should you do this? Well, it depends upon how much you use your scope. If it is running all day, and every day, you probably need to check it over every three months. But this is only an approximate figure, since it makes a difference if it is solid-state or tube, whether it is used in a laboratory-type environment or lugged around in a truck, and so on. You should check it more frequently to begin with, and then decide by experience what is best for your circumstances. You should always check it after making any repair.

Usually your scope will not need complete testing and calibration. Only one or two tests will be needed, and they can be done without completing the entire test procedure. In the following section we'll show how to check each of the main sections, and what to do if adjustment is required. Don't adjust anything unless it is really necessary. In the following procedure the controls are named as in Figure 1.3. If your scope is different, adjust the equivalent control. If its manual gives its own procedure, follow that.

Test Equipment Required

The following instruments, or their equivalent, are required to check out your scope:

An accurate VTVM that reads DC, AC and resistance

An audio oscillator

A signal generator

A time-mark generator, if possible.

Turn On

Before plugging your scope in, measure resistances to ground at the vertical and horizontal inputs (Y and X terminals). In just about all scopes the resistance will be 1 megohm with the AC-GND-DC switch set to DC for the Y input, and the same for the X input with the triggering source set to "X-Y," "horizontal input," or however it is marked. (Some scopes do not have an AC-DC switch for the X input.) In all other positions of the AC-GND-DC switch the resistance reading should be "infinity." If you read 1 megohm in the AC position, suspect the switch or the input coupling capacitor.

It is seldom necessary to measure resistance to ground of the power supplies unless you have made a repair in one of them. The manufacturer's manual should say what resistances to expect.

When turning on your scope for the first time after repair in any power circuit, turn the intensity and one positioning control fully clockwise before applying power. This precaution is to save the CRT screen from damage if something is wrong in the CRT circuit.

Now turn your scope on, and let it warm up for 15 minutes. If you are able to get a small, round, and sharply focused spot with good brilliance, the high voltage power supply can be assumed to be operating properly.

If necessary, adjust FOCUS control at low intensity to obtain a small, round, and sharply focused spot.

If the cathode-ray tube physical position has been changed, check the trace alignment. Set SWEEP TIME switch to 50 MICROSECONDS/DIV. Set the TRIGGERING MODE to AUTO. The CRT trace should be parallel with the graticule. The CRT face should be positioned just behind the filter. (If CRT face and filter touch, Newton rings will occur.)

Power Supplies

The power supplies in your scope are extremely stable so unnecessary adjustments should be avoided.

1. Turn sweep generator off by setting the TRIGGERING SOURCE switch to X-Y.
2. Measure power supply output voltages for normal voltage. See the manufacturer's manual for the proper values.

Trigger Generator

The Trigger Generator must be working in order to adjust the Vertical and Horizontal Amplifiers. Also, these amplifiers must be adjusted before the Sawtooth Generator can be adjusted. To check the Trigger Generator proceed as follows:

TRIGGER SENSITIVITY

Set controls as follows:

VOLTS/DIV (Ch. 1) ------------2m
SOURCE ------------------------------INT
SLOPE -----------------------------+
SEC/DIV -------------------------------50
AC-GND-DC (Ch.1) -----------------DC

With a 10-kHz sine wave fed into Ch.1, decrease the input to the point where triggering becomes unsteady. Vertical deflection should be less than 0.5 div. Check triggering over entire bandwidth.

LINE SYNCHRONIZATION

Set controls as follows:

SOURCE ------------------------------LINE
MODE ---------------------------------AUTO

The normal 60-hertz pickup in the Vertical Amplifier should give a stable pattern.

Vertical Amplifier (Ch.1)

BALANCE

Set controls as follows:

VOLTS/DIV (Ch.1)------------------10
TRIGGERING SOURCE------------X-Y

Turn VAR (Ch.1) fully counter-clockwise and position spot to center with POSITION control. Turn VAR clockwise. Spot should remain centered.

GAIN

Set controls as follows:

VOLTS/DIV (Ch.1)------------------10 m
AC-GND-DC (Ch.1)-----------------GND
SEC/DIV------------------------------50

Connect an audio oscillator to the Y input, and adjust it for a 1-kHz output of 28.3 mV rms as measured on the VTVM. You should observe a signal exactly 8 divisions high on the scope screen. Repeat the above tests for Ch.2.

Horizontal Amplifier

GAIN

Set controls as follows:

TRIGGERING SOURCE------------X-Y
X1-X10---------------------------------X10

Apply 0.35 V rms (as measured on the VTVM) 1-kHz signal to the X input. This should give exactly 10 divisions of deflection.

Sweep Generator

X1 CALIBRATION

Set controls as follows:

SEC/DIV ------------------------------ 10 m
SWEEP EXPANDER ---------------- X1
TRIGGERING SOURCE ------------ INT
TRIGGERING MODE --------------- AUTO

Connect 10-millisecond time markers to input terminal. If a time-mark generator is not available, use 100-hertz sine-wave from audio oscillator (if accurate). The display should show 1 marker or sine-wave per division.

X10 CALIBRATION

Set controls as follows:

SWEEP EXPANDER ---------------- X10

Apply 1-millisecond time markers or 1-kHz sine-wave to Y input terminal. The display should show 1 marker or sine-wave per division.

Time Base Accuracy Check

Set controls as follows:

SEC/DIV ------------------------------ 10μ
SWEEP EXPANDER ---------------- X1
TRIGGERING SOURCE ------------ INT
TRIGGERING MODE --------------- AUTO

Connect 10-microsecond time markers (or 100-kHz signal) to Y input. Each marker or sine-wave should coincide with a vertical graticule line.

Change SEC/DIV to 1μ, and repeat with 1-microsecond time markers (or 1-MHz signal).

Change SEC/DIV to .1m, and repeat with 100-microsecond time markers (or 10-kHz signal).

Change SEC-DIV to 1m, and repeat with 1-millisecond time markers (or 1-kHz signal).

Change SEC-DIV to 10m, and repeat with 10-millisecond time markers (or 100-hertz signal).

Change SEC/DIV to .1, and repeat with 100-millisecond time markers (or 10-hertz signal).

Adjustments

The locations of internal adjustments to correct any out-of-tolerance condition revealed by the foregoing checks are given in the manufacturer's manual. To get at them it will be necessary to remove the scope's outer cover. This is often secured by screws at the rear, and the scope chassis is slid out forward when they have been removed. Follow the manufacturer's instructions in making any adjustments.

When operating your scope with the cover removed be very careful of high voltage, especially around the CRT terminals and power supplies.

Troubleshooting

This section tells you how to isolate the defective stage if something goes wrong with your scope. It doesn't provide detailed information for troubleshooting within the circuits because there is too much variation between different models. The manufacturer's manual for your scope will have the schematic diagrams and other information you need for this.

Before attempting any troubleshooting you should check all controls for proper settings. If you are in doubt about control settings you should review the operating instructions of your scope manual. When you are sure the trouble is in the scope and not the operator you can then proceed to isolate the defective circuit!

Power-supply output voltages should be checked whenever any type of trouble occurs. Sometimes only one scope circuit is malfunctioning, so the last thing you'd suspect is the power supply. But it is possible for an incorrect power-supply voltage to affect one circuit more than others, and valuable time can be saved by checking the power supplies first.

When you've isolated a trouble to a definite circuit, perform a complete visual check of that circuit. This is often the easiest way to spot the trouble. A burnt resistor or a dead tube, for instance, can often be seen at once, and no further testing is necessary. Incidentally, in tube scopes, tube failure is the most likely cause of circuit failure. However, don't rely on a tube tester to tell you whether a tube is good or bad. Check it by substituting another one that is known to be good. If this clears up the problem, you've located the cause. If it does not, return the original tube to its socket, and look elsewhere for the solution.

The troubleshooting procedures that follow are divided into sections according to the symptoms. First find the heading that describes the symptom, then perform the checks given under that heading. Again, we are using the scope illustrated in Figure 1.3 as our model.

No Spot or Trace

No spot or trace on the CRT screen usually means unbalance in the vertical or horizontal amplifier circuits, or trouble in the CRT circuit. First check the beam finder. Depressing this push-button reduces the gain of the vertical and horizontal amplifiers, so an on-screen display is presented regardless of the input signal. The intensity is also brought to a preset level so the trace will be visible in spite of misadjusted controls.

If the beam is positioned off the screen, use the appropriate position controls to attempt to center it. If you can't do this, it means either the vertical or horizontal amplifier is unbalanced. The direction in which the beam-finder trace lies tells you which amplifier is at fault. However, if your CRT has neck pins for the CRT deflection plates, check whether one of these has not lost its lead.

If no beam appears when the beam-finder push-button is depressed, the problem is probably in the CRT circuit. However, to make sure, place a jumper between the vertical or horizontal deflection-plate connections. If no trace appears, the problem must be in the CRT circuit, including the CRT itself.

Waveform Distortion or Insufficient Vertical Deflection

The trouble here must be in the vertical amplifier, including a plug-in if this is a plug-in scope (in the latter case, try another plug-in to determine if it is the plug-in). A common form of waveform distortion results from an improperly-adjusted compensated probe. Check this by substituting another probe. If the cause of the trouble is in the probe, adjust it, using the square-wave output of the PROBE ADJ jack (in some scopes you'd use the calibrator signal).

Insufficient Horizontal Deflection

The cause of the trouble could be in the sweep circuit or the horizontal amplifier. To determine which, set the TRIGGERING SOURCE to X-Y and apply a 3.5-volt rms (10 volts peak to peak) signal to the X input with the variable magnifier set to X1. An approximately 10-division horizontal deflection will demonstrate that the horizontal amplifier is working correctly, so the trouble would have to be in the sweep circuit; otherwise, it would be the horizontal amplifier, of course.

Non-Linear Horizontal Sweep

The linearity of the sweep can easily be checked by connecting the PROBE ADJ output waveform to the Y input, with the SEC/DIV control set to 1m. Adjust the variable and position controls to space the waveform evenly across the CRT screen. If it is non-linear, you won't be able to do this. The trouble could be in the sweep circuit or the horizontal amplifier.

Poor Triggering

If triggering is poor on internal, change to external, connecting the X input to the signal at the Y input. If triggering is now normal, the trouble must be in the trigger pickoff circuit in the vertical amplifier. If it is still poor, you have isolated it to the triggering circuits. Proper operation may be restored by recalibration.

Change of Deflection Sensitivity

If the deflection sensitivity of both the vertical and horizontal amplifiers changes, you can suspect the low-voltage power supplies. Or, it could be caused by a change in CRT sensitivity due to improper output voltages from the high-voltage power supply. Checking the output voltages of both power supplies will reveal the cause.

Repair Procedure

Most of the electronic components in your scope are standard items available locally. When selecting a replacement part you should bear in mind that the physical size and shape of a component may affect its performance in the circuit. You should always install and dress the leads of the new part in exactly the same way as the old part, since many components are mounted in a particular way to reduce or control stray capacitance or inductance.

When replacing a component on a printed circuit board you should remove the board from the scope unless the back of the board is accessible. Use a 15-watt pencil soldering iron with a 1/16-inch wide chisel tip, and electronic grade 60/40 solder. The tip of your iron should be clean and properly tinned for best heat transfer in the shortest time. Do not use a larger iron or you will ruin the bond between the printed circuit and the board.

You'll find the following standard technique to work best:

1. Remove component by cutting its leads near component body.
2. Grip each lead with your needle-nose pliers. Apply tip of soldering iron to connection on back of board, and pull gently to remove lead.
3. When lead comes out it should leave a clean hole. If not, reheat solder and clean out hole with a toothpick or "solder sucker."
4. Clean leads on new component and bend them to correct shape to fit holes. Insert them so that new component seats exactly the same way as old component. If it does not, reheat the connection and gently press the component into place. Remember that some components can be damaged by heat. With these, connect a heat sink such as needle-nose pliers, or a hemostat if you have one, between the component and connection before applying soldering iron.
5. Apply soldering iron to connection at back of board. Use only amount of solder required to form a good electrical connection.
6. Check component (front) side of board to verify that solder has flowed through hole. If lead should also be soldered to conduc-

tor on this side of board, apply tip of iron to connection and apply required amount of solder for a good electrical connection.

7. Clip off excess lead protruding through hole in board.
8. If necessary, clean area around soldered connection with flux remover.

When soldering to a metal terminal, such as a lug on a switch or potentiometer, it is all right to use a larger iron (50 watts, for instance) with a ⅛-inch tip. In fact, you may not be able to manage with a smaller iron, if more than one lead is attached to the lug, because so much heat is absorbed by the additional metal.

Glossary of Terms

accelerating voltage—The cathode-to-viewing-area voltage applied to a cathode-ray tube for the purpose of accelerating the electron beam

alternate display—A means of displaying output signals of two or more channels by switching the channels in sequence.

astigmatism—In the viewing plane of the cathode-ray tube, any deviation of the indicating spot from a circular shape.

attenuator—A device for reducing the amplitude of a signal without deliberately introducing distortion.

automatic triggering—A mode of triggering in which one or more of the triggering circuit controls are preset to conditions suitable for automatically displaying repetitive waveforms. The automatic mode may also provide a recurrent trigger or recurrent sweep in the absence of triggering signals.

bandwidth—Of an oscilloscope, the difference between the upper and lower frequency at which the voltage or current response is .707 (−3 dB) of the response at the reference frequency. Usually, both upper and lower limit frequencies are specified rather than the difference between them. When only one number appears, it is taken as the upper limit.

Note 1: The reference frequency shall be (1) for the lower bandwidth limit, 20 times the limit frequency, and (2) for the upper bandwidth limit, 1/20 the limit frequency. The upper and lower reference frequencies are not required to be the same.

Note 2: This definition assumes the amplitude response to be essentially free of departures from a smooth roll-off characteristic.

Note 3: If the lower bandwidth limit extends to DC, the response at DC shall be equal to the reference frequency, not −3dB from it.

beam finder—A provision for locating the spot when it is not visible.

blanking—Extinguishing of the spot. Retrace blanking is the extinction of the spot during the retrace portion of the sweep waveform. The term does not necessarily imply blanking during the holdoff interval or while waiting for a trigger in a triggered-sweep system.

brightness—The attribute of visual perception in accordance with which an area appears to emit more or less light.

chopped display—A time-sharing method of displaying output signals of two or more channels with a single cathode-ray tube gun, at a rate which is higher than, and not referenced to, the sweep rate.

chopping transient blanking—The process of blanking the indicating spot during the switching periods in chopped display operation.

common-mode signal—The instantaneous algebraic average of two signals applied to a balanced circuit, both signals referred to a common reference.

conventional mode—That mode of operating a storage tube where the display does not store but performs with the usual phosphor luminance and decay.

deflection blanking—Blanking by means of a deflection structure in the cathode-ray tube electron gun which traps the electron beam inside the gun to extinguish the spot, permitting blanking during retrace and between sweeps regardless of intensity setting.

deflection factor—The ratio of the input signal amplitude to the resultant displacement of the indicating spot (for example, volts/division).

delay pickoff—A means of providing an output signal when a ramp has reached an amplitude corresponding to a certain length of time (delay interval) since the start of the ramp. The output signal may be in the form of a pulse, a gate, or simply amplification of that part of the ramp following the pickoff time.

delayed sweep—A sweep that has been delayed either by a predetermined period or by a period determined by an additional independent variable.

dual beam—A multi-trace cathode-ray tube which produces two separate electron beams that may be individually or jointly controlled. In contrast to dual-trace operation, waveforms are not chopped or switched; they are generally brighter and have minimum phase relationship error.

dual-trace—A multi-trace operation in which a single beam in a cathode-ray tube is shared by two signal channels. See *alternate display, chopped display* and *multi-trace*.

focus—Maximum convergence of the electron beam manifested by minimum spot size on the phosphor screen.

Gaussian response—A particular frequency response characteristic following the curve $y(f) = e^{-af^2}$. Typically, the frequency response approach by an amplifier having good transient response characteristics.

geometry—The degree to which a cathode-ray tube can accurately display a rectilinear pattern. Generally associated with properties of a cathode-ray tube electrode or its associated control.

graticule—A scale for measurement of quantities displayed on the cathode-ray tube of an oscilloscope.

input RC characteristics—The DC resistance and parallel capacitance to ground present at the input of an oscilloscope.

intensity modulation—The process and (or) effect of varying the electron-beam current in a cathode-ray tube, resulting in varying brightness or luminance of the trace.

internal graticule—A graticule whose rulings are a permanent part of the inner surface of the cathode-ray-tube faceplate.

jitter—An aberration of a repetitive display indicating instability of the signal or of the oscilloscope. May be random or periodic, and is usually associated with the time axis.

magnified sweep—A sweep whose time per division has been decreased by amplification of the sweep waveform rather than by changing the time constants used to generate it.

mixed sweep—In a system having both a delaying sweep and a delayed sweep, a means of displaying the delaying sweep to the point of delay pickoff and displaying the delayed sweep beyond that point.

multi-trace—A mode of operation in which a single beam in a cathode-ray tube is shared by two or more signal channels. See *dual-trace, alternate display* and *chopped display*.

resolution—A measure of the total number on trace lines discernible along the coordinate axes, bounded by the extremities of the graticule or other specific limits.

risetime—In the display of a step function, the interval between the time at which the amplitude first reaches specified lower and upper limits. These limits shall be 10 percent and 90 percent of the nominal or final amplitude of the step, unless otherwise stated.

roll-off—A gradually increasing loss or attenuation with increase or decrease of frequency beyond the substantially flat portion of the amplitude-frequency response characteristic of a system or transducer.

signal delay—In an oscilloscope, the time required for a signal to be transmitted through a channel or portion of a channel. The time is always finite, may be undesired, or may be purposely introduced as in a delay line.

tangential noise measurement—A procedure to determine displayed noise wherein a flat-top pulse or square-wave input signal is adjusted in amplitude until the two traces (or portions of two traces), thus produced, appear to be immediately adjacent or contiguous. Measurement of the resulting signal amplitude determines a noise value which correlates closely with the value interpreted by the eye from a sampling display and is called the "tangential noise value."

trigger—A pulse used to initiate some function (for example, a triggered sweep or delay ramp).

unblanking—Turning on of the cathode-ray-tube beam.

How to Use a Delaying Time Base

A delaying time base is provided in some scopes for the purpose of expanding a selected portion of the display. Such scopes, therefore, have *two* time bases, usually called A and B. The most noticeable difference between a delaying time base and a standard time base is the "delay time multiplier" control.

This control sets the threshold level of a comparator in the delaying time base B. When the sawtooth sweep voltage in this time base reaches the threshold level the comparator generates a trigger pulse that starts the sweep in the other time base A.

The scope horizontal mode control can be set for using A or B independently, when they operate like normal time bases, or in combination. In the combination setting called "B intensified by A" the display will look like the upper part of Figure A.1. Here the sweep A is superimposed on the sweep B, causing a portion to be brightened or "intensified." In this case the speed of sweep A was set to be ten times as fast as that of sweep B, so the brightened portion is only one-tenth of the trace. Its length could be varied by changing the sweep rate.

The delay time multiplier control adjusts the time interval between the start of the B sweep and the start of the A sweep. Rotating this control causes the brightened portion to move back and forth on the trace according to the direction of rotation. This enables you to position it so it intensifies any part of the trace you are especially interested in. Having done this, you switch the horizontal mode control to the other combination position "A delayed," and the brightened portion expands to the full width of the graticule. This is illustrated in the lower part of Figure A.1. You can see that this

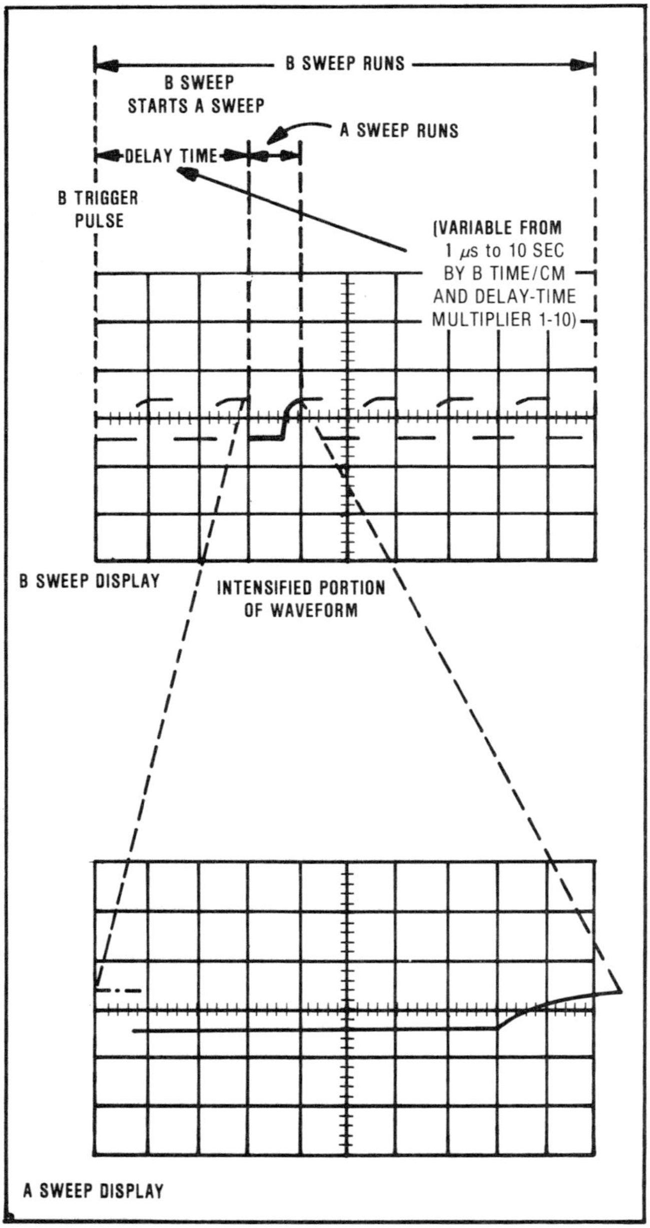

Figure A.1 What the delaying time base does. (*Reprinted by permission of Tektronix, Inc. All rights reserved.*)

gives a result similar to a magnified sweep, but it has better resolution and stability, and greater ratios of effective magnification.

The delay time multiplier control can also be used to measure the time interval between two points on the display. With the horizontal mode control set to "B intensified by A" a small portion of the display will be brightened, as we've already seen. By rotating the delay time multiplier control you can set this brightened portion at the first point. Make a note of the reading on the dial of the delay time multiplier, then move the brightened portion to the second point by rotating the control. Make a note of the new reading on the dial. The difference between the two readings, multiplied by the B sweep time base, gives you the time interval between the two points. This is a much more accurate method than using the graticule.

INDEX